Breathing Wall
— Architecture Practical Cases of Vertical Greening

会呼吸的墙
——建筑立体绿化实例

君誉文化　策划
高迪国际出版（香港）有限公司　编
贾艳春　米晓丽　译

大连理工大学出版社

图书在版编目(CIP)数据

会呼吸的墙:建筑立体绿化实例/高迪国际出版(香港)有限公司编;贾艳春,米晓丽译.—大连:大连理工大学出版社,2016.9
　ISBN 978-7-5685-0440-9

Ⅰ.①会… Ⅱ.①高… ②贾… ③米… Ⅲ.①建筑物—墙—绿化 Ⅳ.①TU985.1

中国版本图书馆CIP数据核字(2016)第159970号

出版发行:大连理工大学出版社
　　　　　(地址:大连市软件园路80号　邮编:116023)
印　　刷:上海锦良印刷厂
幅面尺寸:235 mm×310 mm
印　　张:17.5
插　　页:4
出版时间:2016年9月第1版
印刷时间:2016年9月第1次印刷
责任编辑:裘美倩
责任校对:仲　仁
策　　划:君誉文化
封面设计:高迪国际

ISBN 978-7-5685-0440-9
定　　价:328.00元

电　话:0411-84708842
传　真:0411-84701466
邮　购:0411-84708943
E-mail:designbookdutp@gmail.com
URL:http://www.dutp.cn

如有质量问题请联系出版中心:0411-84709043　0411-84709246

FOREWORD
前言

城市绿化景观发展新方向
——建筑墙体绿化

社会经济的高速发展促使城市建设不断扩大，城市内高楼林立，绿地面积越来越少，严重破坏了城市的生态环境。生态环境破坏最为严重的是生态系统的消失：绿地、森林和湿地被高楼大厦、工厂厂房、经济开发区和公路所取代，城市污染物无法得到有效的降解和消除。城市建筑的外表面，拥有大面积无须采光的墙体，为墙体绿化提供了大量空间。

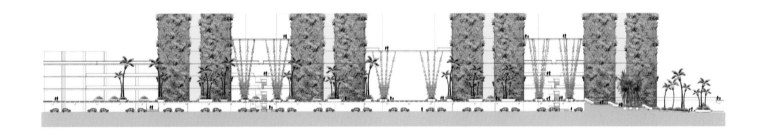

认识绿墙

绿墙，是通过在建筑物室内外墙壁栽植各种植物来绿化、美化墙体，营造一种局部的自然生态氛围。泛指用攀援植物或其他植物装饰建筑垂直面或各种围墙的一种垂直绿化形式，以达到美化和维护生态的目的。这种生长着的"墙"叫生态绿墙，目前欧美国家、日本、加拿大均有大量运用。生态绿墙采用先进绿墙设计与应用技术，系统超轻超薄，无须骨架安装，占用空间小，可以在任何规则与弧度的墙体上施工；智能自动灌溉系统，现代无土栽培技术，无须管理，植物存活时间长，而且可以根据客户的需求进行各种艺术图案的处理。

绿墙的优势

适用面广

植物绿墙可以用于建筑室内外的各个地方，除了美化、净化环境以外，还可以承担起许多实际功能，比如作为背景墙、艺术隔断、植物壁画等。

保护建筑

一般绿化的面积和形状都可人为地控制，可让绿化植物成图案式覆于墙面，形成景观。通过形体与色彩艳丽的植物使线条生硬、色彩灰暗的建筑材料变得自然柔和。

节约能源

研究表明，当气温超过 34℃ 时，有屋顶及墙体绿化的房间，气温要下降 2℃～5℃，空调负荷减少 15% 以上，真正符合节能环保的现代理念。

生态保护

就拿一个几十平方米的房间来说，一面绿墙就是一个微型的生态系统，这里有几百株绿色植物，能够隔热降温，降噪除尘，调节室内湿度，吸附空气中的有害物质，净化空气，增加大量负氧离子含量，从而改善环境。

造价低廉

用于垂直绿化的植物具有极强的生命力、易繁殖蔓延、环境适应能力强；对土壤、水、肥等生存环境要求不高且基本不需要整形修剪，墙体绿化造价低、养护方便。

绿墙构成要素

植栽选种： 植物的选择要依据绿墙设置的地域、朝向及设计效果，一般更换周期长的绿墙多选用生长旺盛的植物。

栽培介质： 需要综合考虑绿墙的整体设计及其搭配的系统结构，通常按一定比例或全部使用轻介质材料。

栽培容器： 因各系统设计不同而有很大差异。

灌溉系统： 根据各系统结构特点配置不同的灌溉系统，有喷雾、喷灌、滴灌等系统。

结构系统： 绿墙结构系统的设置需要考虑结构组件取得的难易、花墙设计高度及墙体绿化总数相加后的平均单位重量，其造价通常与以上要件成正比。

维护保养： 绿墙的维护保养是绿墙能否持续的关键。一般来说，好的系统是低度维护保养、简单操作控制、成本相对较低，而越精密的控制越容易出现问题，包括设备、人员素质等，成本也就不容易掌控。

绿墙的植物选择

由于建筑墙面的日照、温度、空气成分、风力等生态因子与地面明显不同。不同楼层高度及不同地区、不同条件下生态因子也不同，因此要根据屋顶的生态因子的要求选择植物，以便达到预期的效果。墙面绿化植物的选择必须从建筑墙体的环境出发，首先考虑满足植物生长的基本要求，然后才能考虑植物配置艺术。

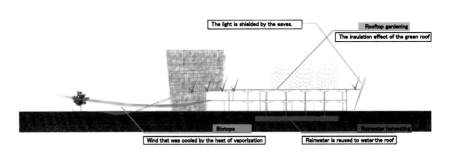

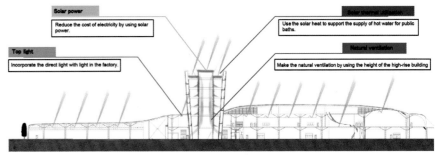

选择耐旱、抗寒性强的植物

由于建筑墙面夏季气温高、风大、土层保湿性能差，冬季则保温性差，所以应选择耐干旱、抗寒性强的植物，还要考虑墙面特殊地理环境和承重的要求。

选择抗风、不易倒伏的植物

建筑墙面上风力比地面大，特别是雨季或有台风来临时，风雨交加对植物的生存危害最大，加上墙面种植层薄，应选择一些抗风、不易倒伏的植物。

选择以常绿为主、冬季能越冬的植物

在建筑墙体上进行绿化的目的是增大城市的绿化面积，美化"第五立面"，植物选择应以常绿为主，宜用叶形和株形秀丽的品种，还可适当栽植一些色叶植物或布置一些盆栽时令花卉，营造四季有景的氛围。

选择耐瘠薄的浅根性植物

建筑墙体大部分为全日照直射，光照强度大，植物应尽量选用阳性植物。建筑墙体的种植层较薄，为了防止根系对建筑结构的侵蚀，应尽量选择浅根系及耐瘠薄的植物种类。

选用乡土植物

尽量选用乡土植物，适当引种绿化新品种。乡土植物对当地的气候有高度的适应性，在环境相对恶劣的时候，选用乡土植物有事半功倍之效。

地被植物

波斯顿蕨

拉丁学名：*Nephrolepis exaltata*
科　　属：肾蕨科肾蕨属
形态特征：多年生常绿蕨类草本植物。根茎直立，有匍匐茎。叶丛生，长可达 60 cm 以上，具细长复叶，叶片展开后下垂。
生态习性：性喜温暖、湿润及半阴环境，又喜通风，忌酷热。
园林用途：下垂状的蕨类观叶植物，适宜盆栽于室内吊挂观赏、盆栽及垂直绿化，其匍匐枝剪下可用作装饰配置材料。

垂盆草

拉丁学名：*Sedum sarmentosum* Bunge
科　　属：景天科景天属
形态特征：多年生肉质草本，不育枝匍匐生根，结实枝直立，花淡黄色，花期 5～6 月，果期 7～8 月。
生态习性：耐干旱、耐高温，抗寒性强，耐湿、耐盐碱、耐贫瘠。抗病虫害能力强，生长速度快，繁殖容易，不用修剪。
园林用途：绿叶观赏期达 8～9 个月。草姿美，色绿如翡翠，颇为整齐壮观；花色金黄鲜艳，观赏价值高。

羽衣甘蓝

拉丁学名：*Brassica oleracea* var. *acephala* f. *tricolor*
科　　属：十字花科芸薹属
形态特征：植株高大，根系发达，茎短缩，密生叶片。叶片肥厚，倒卵形，深度波状皱褶，呈鸟羽状。花序总状。
生态习性：喜冷凉温和气候，耐寒性很强，成株在我国北方地区冬季露地栽培能经受短时几十次霜冻而不枯萎。
园林用途：在华东地带为冬季花坛的重要材料。其观赏期长，叶色极为鲜艳，在公园、街头、花坛常见用羽衣甘蓝镶边和组成各种美丽的图案，用于布置花坛，具有很高的观赏效果。

黑麦草

拉丁学名： *Lolium perenne*

科　　属： 禾本科黑麦草属

形态特征： 一年生或多年生草本。叶在芽中呈折叠状，叶鞘光滑，叶耳细小，叶舌短而不明显。穗状花序。

生态习性： 具有极强的适应性、抗寒等特点。喜温和、湿润、凉爽气候，最适生长温度为 20℃ ~ 25℃。

园林用途： 多年生黑麦草是一草多用的优良牧草。由于其根系发达，生长迅速，耕地种植可增加种植地的土壤有机质，改善种植地土壤的物理结构；坡地种植，可护坡固土，防止土壤侵蚀，减少水土流失。

佛甲草

拉丁学名： *Sedum Lineare* Thunb

科　　属： 景天科景天属

形态特征： 多年生草本，无毛。茎高 10 ~ 20 cm。3 叶轮生，叶线形。花序聚伞状，顶生，花期 4 ~ 5 月，果期 6 ~ 7 月。

生态习性： 适应性极强，不择土壤，耐寒力极强。属多浆植物，含水量极高，其叶、茎表皮角质层具有防止水分蒸发的特性。

园林用途： 佛甲草种植在屋顶上，可以生长在较薄的基质上（3 ~ 5 cm 厚），而且能抗高温。夏天屋顶温度高达 60℃，它也能承受，基本不用浇水。而在冬季寒冷的北方，严寒时基质冻结，佛甲草呈休眠状态；开春后气温回升，很快又能萌发。

非洲凤仙

拉丁学名： *Impatiens wallerana*

科　　属： 凤仙花科凤仙花属

形态特征： 多年生草本。茎多分枝，光滑，节间膨大。叶卵形，边缘钝锯齿状。花腋生，1 ~ 3 朵，花形扁平，花色丰富。

生态习性： 不耐高温和烈日暴晒。喜温暖湿润和阳光充足的环境，土壤宜用疏松、肥沃和排水良好的腐叶土或泥炭土。

园林用途： 在欧美的草本花卉应用中，非洲凤仙花排在第一位。它提供了一个广泛的亮丽颜色系列且长势旺盛，管理简单。适于盘盒容器、吊篮、花墙、窗盒和阳台栽培。

藤本植物

凌霄

拉丁学名：*Campsis grandiflora*
科　　属：紫葳科凌霄属
形态特征：落叶木质藤本。羽状复叶对生。花橙红色，由三出聚伞花序集成稀疏顶生圆锥花丛。蒴果。花期6~8月。
生态习性：喜充足阳光，也耐半阴。耐寒、耐旱、耐瘠薄，病虫害较少。以排水良好、疏松的中性土壤为宜，忌酸性土。
园林用途：凌霄生性强健，枝繁叶茂，是廊架绿化的上好植物。

常春藤

拉丁学名：*Hedera nepalensis* K,Koch var.sin ensis (Tobl.) Rehd
科　　属：五加科常春藤属
形态特征：常绿攀缘灌木，有气生根。叶为单叶。伞形花序单个顶生，果实球形。种子卵圆形。
生态习性：在温暖湿润的气候条件下生长良好，不耐寒。对土壤要求不严，喜湿润、疏松、肥沃的土壤，不耐盐碱。
园林用途：在庭院中可用以攀缘假山、岩石，或在建筑阴面做垂直绿化材料。不仅可达到绿化、美化效果，同时也发挥着增氧、降温、减尘、减小噪声等作用，是藤本类绿化植物中用得最多的材料之一。

金银花

拉丁学名：*Lonicera japonica*
科　　属：忍冬科忍冬属
形态特征：常绿木质藤本植物，多年生，是优美的观赏花卉。茎褐，幼嫩枝条绿色，有毛。叶卵圆形，有短柄。花期4~6月。
生态习性：适应性强，喜阳、耐阴，耐寒性强，耐干旱和水湿，对土壤要求不严，根系繁密发达，萌蘖性强，茎蔓着地即能生根。
园林用途：由于匍匐生长能力比攀缘生长能力强，故适合做绿化矮墙；亦可以利用其缠绕能力制作花廊、花架、花栏、花柱以及缠绕假山石等。

扶芳藤

拉丁学名：*Euonymus fortuni*

科　　属：卫矛科卫矛属

形态特征：攀缘藤本。叶对生卵形，革质，浓绿色。枝条上有细密微突气孔，能随处生根。5～6月开花，聚伞花序。

生态习性：耐阴，不喜阳光直射。喜湿润，夏季盆栽植株需早晚浇水，冬季减少浇水量。喜温暖，较耐寒，江淮地区可露地越冬。

园林用途：庭院中常见的地面覆盖植物，点缀墙角、山石、老树等，都极为出色。其攀缘能力不强，不适宜作立体绿化（变种爬行卫矛则可）。如对植株加以整形，使之成悬崖式盆景，置于书桌、几架上，可给居室增加绿意。

爬山虎

拉丁学名：*Parthenocissus tricuspidata*

科　　属：葡萄科地锦属

形态特征：爬山虎属多年生大型落叶木质藤本植物。藤茎可长达18 m。夏季开花，花小，成簇不显，与叶对生。

生态习性：适应性强，性喜阴湿环境，但不怕强光，耐寒，耐旱，耐贫瘠，气候适应性广泛。

园林用途：爬山虎是最常用也是最理想的攀缘植物，它依靠吸盘沿着墙壁往上爬。种植的时间长了，密集的绿叶覆盖了建筑物的外墙，就像穿上了绿装。

紫藤

拉丁学名：*Wisteria sinensis* (Sims)Sweet

科　　属：豆科紫藤属

形态特征：落叶木质大藤本。奇数羽状复叶。4月开花，花蓝紫色，总状花序下垂，长15～30 cm，有芳香。

生态习性：性喜光，略耐阴，耐干旱，忌水湿。生长迅速，寿命长，深根性，适应能力强。

园林用途：优良的观花藤木植物，一般应用于园林棚架，春季紫花烂漫，别有情趣，适栽于湖畔、池边、假山、石坊等处，具有独特风格，在绿化中已得到广泛应用，尤其在立体绿化中发挥着举足轻重的作用。

绿墙新类型

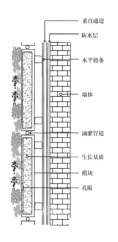

模块式

利用模块化构件种植植物实现墙面绿化的形式。绿化模块是由种植构件（盒）、种植基质和植物三部分组成的。植物生长须具备养分、水和空气三要素，作为容器的种植构件，需满足植物生长的必备条件，如固定植物的根系、蓄水、排水、空气循环以及和建筑之间的悬挂固定等要求。施工方法是将预先栽培养护好的植物根据方形、菱形、圆形等几何单体植物模块构件，通过合理的搭接或绑缚固定在不锈钢等骨架上，形成各种形状构图和景观效果。种植形式为在模块中预先栽培，植物则按设计的图案要求预先栽培养护数月。该类型的特点是，绿化植物寿命较长，适用于大面积高难度的墙面绿化，墙面景观营造效果好，适宜浇灌和微灌。

铺贴式

在墙面直接铺贴生长基质与植物组成的栽培平面系统，或者是采用喷播技术在墙面形成一个种植系统。施工方法是直接铺贴生长基质与植物组成的栽培系统或喷播。该类型的特点是可以将植物在墙体上自由设计或进行图案组合；直接附加在墙面，无须另外做钢架，并通过自来水和雨水浇灌，降低建造成本；系统总厚度薄，只有 10 ~ 15 cm，并且还具有防水阻根功能，有利于维护建筑物，延长其寿命；易施工，效果好等。

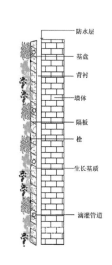

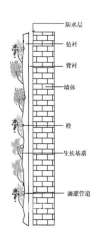

布袋式

在铺贴式墙面绿化系统基础上发展起来的一种工艺系统。这一工艺首先在做好防水处理的墙面上直接铺设软性植物生长载体，比如毛毡、椰丝纤维、无纺布等，然后在这些载体上缝制装填有植物生长基材的布袋，最后在布袋内种植植物实现墙面绿化。该类型的特点是，施工简便、造价较低、透光透气性好，能充分利用雨水浇灌，适宜在大面积的边坡治理及水土保持壁面上应用。

注：本段内容论文出处为《墙面绿化新技术浅析》，作者为何国强（深圳大学园艺中心）；彭坚（深圳职业技术学院）；刘春常、许建新、黄东光（深圳市铁汉生态环境股份有限公司）。

摆花式

在由不锈钢、钢筋混凝土或其他材料等做成的垂直面架中安装盆花或直接在建筑墙面上安装人工基盘实现墙面绿化的形式，在人工支架的基础上，装上各种各样的栽培基质穴盘，穴盘有卡盆式、包囊式、箱式、嵌入式等不同种类。种植形式一般为盆栽摆设。该类型的特点是，适用于临时墙面绿化或立柱式花坛造景，一般使用滴灌或雾喷，通常是在人工基盘上接入微灌设施以减轻管护压力。

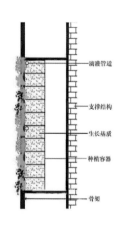

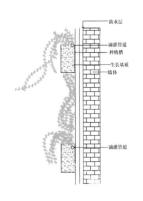

攀缘或垂吊式

这是目前最普遍的一种垂直绿化形式，多用于边坡绿化、墙面绿化等，即在墙面种植攀爬或垂吊的藤本植物，如种植爬山虎、络石、常春藤、扶芳藤、绿萝等。该类型的特点是：简便易行、变化灵活、造价较低、透光透气性好，能利用雨水浇灌，绿化植物存活时间较长，管理成本低。

V 形板槽式

在墙面上按一定的距离安装 V 形板槽，在板槽内填装轻质的种植基质，再在基质上种植各种植物。这类绿化形式的特点是：施工简便灵活、造价低、适宜种植的植物种类多，兼备摆花式、攀爬式及垂吊式墙面绿化的优点。适用的植物种类较多，可组合栽植灌木、花草或蔓生性强的攀爬或垂吊的植物，既可浇灌也可微灌。

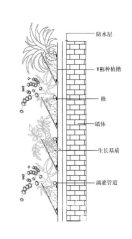

结语

目前，国内很多城市都已开始实施建筑墙体绿化，其建设形式也在不断更新。墙体绿化作为垂直绿化的一种类型，不仅可以满足建筑结构、气候环境、植物材料等不同建设条件要求，还可以在更大范围内发挥生态效应，充分体现其先进性，同时能营造更为绚丽多姿的景观，让城市环境更优美。

Veera Sekaran
Managing Director of Greenology

1 PREFACE
序言 1

Vertical greenery is a delicate balance – a marriage between form and function, pragmatism and imagination, nature and technology, biology and chemistry, precision and free-growth, to produce a multitude range of beautiful oxygen-producing walls cladded in a multitude of plant species. While cities and urban spaces continue to grow, vertical greenery is fast growing and evolving as one of the significant antidotes to urban planning challenges for now and many years to come. Viewed from many aspects – health and wellbeing, environmental care, human development, economics, aesthetics and lifestyle – all these nod in agreement that green walls are here to stay for a long time to come. People realise that nature brings comfort, solace, cleanliness and purity. Plants are a life-giving source that combats and addresses the contaminated air that we breathe in, artificially generated urban heat and dirt, modern-day stressful living, and disengagement with our natural surroundings and natural self.

With the correct, affordable technology, low in maintenance and sustainable for the long time, some green walls can support a large variety of plant species. Vertical ecosystems can enhance the biodiversity in the urban environment by encouraging flora and fauna to survive together with people, up close. This vital symbiosis contributes to the well-being of humans in the urban environment – provision of clean air, food, mood enhancement through visual, audio and colour variations of the living green walls, reuse of grey water, etc. A vertical garden should have a variety of plants thriving together symbiotically to ensure a sustainable garden. Native and locally adapted plants are chosen based on adaptability to grow vertically, sunlight requirement and aesthetic factors in designing a green wall. Therefore, designing a vertical garden is both a science and an art – to create an impressive garden that is beneficial to the vertical ecosystem.

A green revolution has taken place and there will only be more exciting things to come in the frontier of vertical greenery. What is interesting is to see how green walls will continue to evolve – already we have been enjoying lower costs of vertical greenery as the technology gains momentum and as the scale of economics tips to the advantage of the consumer with an increasing pool of users.

What was once considered a luxury and novelty in greening will continue to gain more mainstream support and believers. The momentum and growing support of like-minded people who believe in and lobby for green walls are natural catalysts to the greening industry, a blessing to all who cherish nature. As a more educated society progress, it will see the many strong benefits green walls bring to mankind grounded by scientific proof. Hence, we can expect more vertical gardens sprawling around the cities, greening our environment.

垂直绿化是一个微妙的平衡，通过形式与功能、实用性与想象力、自然与技术、生物与化学、精密度与自由生长之间的密切结合，来打造大范围外观漂亮、由众多植物种类构成，并具有氧气制造功能的墙壁。随着城市和城市空间的不断发展，垂直绿化发展迅速，成为现在及未来许多年城市发展挑战的一剂良药。从健康、福利、环境、人类发展、经济、美学和生活方式等很多方面来看，绿墙结构都将长期存在。人们意识到自然能够带来舒适、安慰、清洁和纯净。植物是生命的源泉，它与我们吸入的污染空气、人为产生的城市高温和污垢以及当代紧张并与大自然相脱离的生活状况相抗衡。

使用恰当以及可承受的技术，一些绿墙维护成本低，持续时间久，可以支撑很多种类的植物。垂直生态系统通过鼓励动植物群与人类一起零距离接触，来加强城市环境的生物多样性。这个重要的共生关系给城市中生存的人类在环境方面带来福利，比如通过植生墙在视觉、声音和颜色上的变化以及灰水的再利用等方法，给人们带来新鲜空气、食物以及心情的变化。垂直花园应该拥有各种各样的植物，共生繁盛，以此确保花园的可持续性。基于适应性，选择适合本国和当地生长的植物，垂直种植，在设计中考虑光照和美学因素。因此，垂直花园设计既是一门科学又是一门艺术，打造一个引人入胜的花园对垂直生态系统来说益处颇多。

绿色革命已经发生，垂直绿化技术的发展前沿将更加有趣。现在有趣的是，绿墙设计将如何持续发展，随着科技的加速发展和经济规模向越来越多的消费者利益方面倾斜，我们已经享受到了低成本的垂直绿化。

垂直绿化曾经被认为是在绿化方面的奢侈品和新颖事物，现在其将继续得到更多的支持与信赖。志趣相投的人们不断支持，他们对绿墙设计的信任和游说犹如绿化产业自然的催化剂，对珍爱大自然的人们是一个福音。随着社会教育的发展，以科学为基石，我们将看到绿墙给人类带来更多的利益。因此，我们期待更多的垂直花园在城市中蔓生，绿化我们的环境。

2 PREFACE
序言 2

Michael Leong
Architect of SAA Architects

Sky-rise greenery in Singapore is not new. Ever since people started living in high-rise public housing, the phenomenon was born in the humble form of potted plants along corridors and balconies. Biophilia, or the affinity to nature, is the driving force behind this instinctive behavior. It has a strong effect on the urban environment, as it softens the city and brings nature back to the daily lives of urbanites.

Over the years, vertical greenery have been developed and applied on buildings like an architectural system, and becoming just as important as curtain walls or roof tiles in some cases. This 'covering' of buildings by greenery is a step towards the idea of 'City in a Garden', where hard façade surfaces and materials are exchanged for soft living plants.

Buildings also start to become microcosms of the city when green spaces are incorporated onto them. They are often communal green pockets right at the door steps of the residents, especially in mixed developments, where extensive social gardens can be incorporated. In a fast-paced city such as Singapore, people are increasingly short of time for leisure and travel, making such green spaces more desirable and valuable.

If this trend of sky-rise greenery continues to intensify, it can even create a harmonizing effect among buildings in Singapore. If developers and architects constantly address the harsh tropical climate with sky-rise greenery, it can even become a Singaporean architectural language, like how traditional architecture of Japanese buildings deal with threats of nature. This would mean that we need to treat greenery as an intrinsic part of building design, almost as non-negotiable as waterproofing. We would also need to design planting features as we design facades; making provisions for structural loading, irrigation, exposure to sunlight.

Architects often integrate different façade systems to attain an overall desired performance, such as the combination of low-E glass for heat gain reductions with floating glass for spandrels. In a similar fashion, green walls must also be designed to work optimally. For instance we worked with landscape architects to use different species of plants within a single wall to attain an overall desired result. Another challenge is that we are working with living objects, so a building must still be aesthetically pleasing should the greenery fail to establish during certain seasons of the year. In that vein, structures supporting greenery should be part of the architecture, so they do not appear as temporary works when seen without greenery.

Good planning is vital to making vertical greenery systems work well with a building. This includes planning for the best locations where it will add most value, planning for its nutritional needs so that it will thrive, and planning for the possibility of a failure to establish. When we design sky-rise greenery we are working with living things, hence we need to design for its life, growth and death.

在新加坡，摩天大楼上种植绿色植物并不新鲜。自从在高层建筑中开始生活，人们就已将盆栽的植物放在走廊和阳台上了。生物恋，或对大自然的亲和力，是这种本能行为背后的推动力量。它对城市环境有强大的影响力，因为它柔化了城市，为生活在都市的人们的日常生活带来自然的绿意。

多年来，人们开发了垂直绿色植物，并将其应用到建筑大楼上，在某些项目中，这些垂直植物变得像幕墙和屋瓦一样重要。这种绿色"覆盖"是迈向"花园城市"的一步，在这里，坚硬的立面和材料被换成了柔软的、鲜活的植物。

绿色空间被整合进建筑内，大楼成了城市的缩影。它们成为居民区门阶处的公共绿色空间，尤其是在混合建筑区，宽阔的公共花园与建筑融合在一起。在一个生活节奏很快的城市，如新加坡，人们休闲和旅游的时间越来越少，这样的绿色空间就更让人向往，更具有价值。

如果这种对高层建筑进行绿化的趋势得到强化，那么它将在新加坡的建筑群中营造一种和谐的效果。如果开发商和建筑设计师继续用高层绿化来解决严酷的热带气候问题，它将可能成为新加坡的建筑语言，正如传统的日本建筑强调如何应对自然威胁一样。这将意味着，我们需要将绿色植物作为建筑设计的内在组成部分，和防水工程一样重要，这一点毋庸置疑。当我们设计立面时，我们也要根据植物的特色进行设计；规定好结构承载、灌溉和日照。

建筑设计师经常将不同的立面系统整合在一起，以获得整体所需性能，如将可减少吸热的低辐射镀膜玻璃与拱肩的浮法玻璃结合起来。同样地，我们设计的绿墙也必须达到最优工作效果。比如，我们与景观设计师一起在单一墙面使用不同种类的植物，来达到整体的预期效果。另一个挑战是，我们的工作对象是活的植物，因此一栋大楼一定要有美感，即使绿色植物在一年中的某些季节不能为人们提供视觉享受。同样地，支撑绿色植物的结构也应是建筑的一部分，这样即使没有绿色植物的时候，这些结构也不会看起来像是临时设施。

要让垂直绿色系统与建筑良好运行，良好的规划至关重要，包括选择最佳地点，这个地点能最大限度地增加价值；为植物进行营养需求的规划，这样植物能茁壮生长；还有对失败可能性的准备。当我们设计摩天大楼的绿色植物覆盖时，我们的工作对象是活的植物，因此，我们需要为它的生命、生长和死亡进行设计规划。

CONTENTS 目录

Institute of Technical Education
工艺教育学院
020

School of the Arts
艺术学院
030

Jardin
Jardin 项目
038

Hansar
汉莎项目
044

48 North Canal Road
北运河路 48 号办公楼
050

Double Bay Residences
双湾雅居绿化项目
056

Shaw Centre
邵氏大厦
062

The Heeren
麒麟大厦
068

Changi Airport Terminal 3 Interior Landscape
新加坡樟宜机场三号航站楼室内景观
072

158 Cecil Street
158 丝丝街项目
078

Jem®
Jem® 购物中心
090

| 096 | Osaka-Marubiru
大阪 Marubiru 项目 |

| 102 | Social Housing Croix Nivert
克罗斯克尼维特社会住房 |

| 110 | Foundation for Polish Science Headquarters
波兰科学基金会总部项目 |

| 116 | Pasona HQ
保圣那集团总部项目 |

| 122 | The New Construction of the Restaurant and Fitness Center in the Headquarters Building of "CSC"
中国台湾"中钢"总部大楼餐厅及健身中心新建工程 |

| 126 | THSR Branch of Jungli City Kindergarten + Ching Chih Children's Park
中坜市立幼儿园高铁分班 + 青芝儿童公园 |

| 132 | Newton Suites
牛顿套房 |

| 140 | Factory on the Earth
地球工厂 |

| 146 | Medibank Building
墨尔本医疗福利办公大楼 |

| 152 | The Picardy Regional Chamber of Commerce and Industry
皮卡地区商会 |

| 160 | Sportplaza Mercator, Amsterdam, The Netherlands
荷兰阿姆斯特丹墨卡托体育馆 |

NEX Shopping Center
NEX 购物中心　　168

Ocean Financial Center
海洋金融中心　　172

Gramercy Sky park
葛莱美西空中公园　　180

LG Arena
LG 广场　　184

Maputo Commercial Building
马普托商业大厦　　190

Stay
Stay 餐厅　　194

Westfield Living Wall
韦斯特菲尔德购物中心垂直绿墙设计　　198

Zentro Office Building and Commercial
Zentro 商用办公大楼　　208

Angie Fowler Adolescent & Young Adult Cancer Institute
Angie Fowler 青少年及年轻人癌症研究所项目　　214

BB House
BB 大楼项目　　220

Firma Casa
Firma Casa 商店　　224

| 230 | Vallarta House
巴亚尔塔住宅区 |

| 236 | Restructuration of the Brussels Regional Council
布鲁塞尔区政府重建工程 |

| 242 | Green Side-wall
绿色侧墙 |

| 248 | Fashion Valley Mall
时尚谷购物中心 |

| 252 | Illura Apartments
Illura 公寓 |

| 256 | Jindrišská 16
Jindrišská 16 项目 |

| 260 | Qianhai, Vanke
万科前海 |

| 266 | Refurbish of TYJ Office Building
深圳桃源居办公楼改造 |

| 274 | INDEX
索引 |

Location
Singapore

Institute of Technical Education

工艺教育学院

Landscape Architect
Grant Associates

Grant Associates was a part of the winning team with RSP to design and implement the new Headquarters and College Central for the Institute of Technical Education in Singapore. Grant Associates was commissioned to undertake the concept design services through to detailed production information.

College Central is the third and final ITE College, specialising as a college of creativity and innovation. The building and landscape are closely integrated and built up of various layers to ensure the spatial efficiency of the campus. The ground floor plane supports the access, car parking and service requirements whilst the first floor level known as the "Spine" provides a key orientation and social space which connects to several aerial walkways and the terraces to the 12 educational blocks.

The Spine hosts a series of gardens and planted "plinths" containing specimen trees and colourful "wire trees". The gardens and plinths create distinct spaces and key circulation nodes around the vertically stacked aerial walkways. A continuous water feature flows from the source garden down to the arrival garden, creating cool seating areas within the shade of the pod gardens. Connecting into the Spine are a series of retreat terraces and sunken gardens, providing a range of spaces for students and staff to utilise that connect out into the wider campus landscape.

The campus has been designed for a student population of 10,400 and 850 staff. The new ITE Headquarters is carefully integrated into the campus, accommodating a total of 550 staff along with conference facilities.

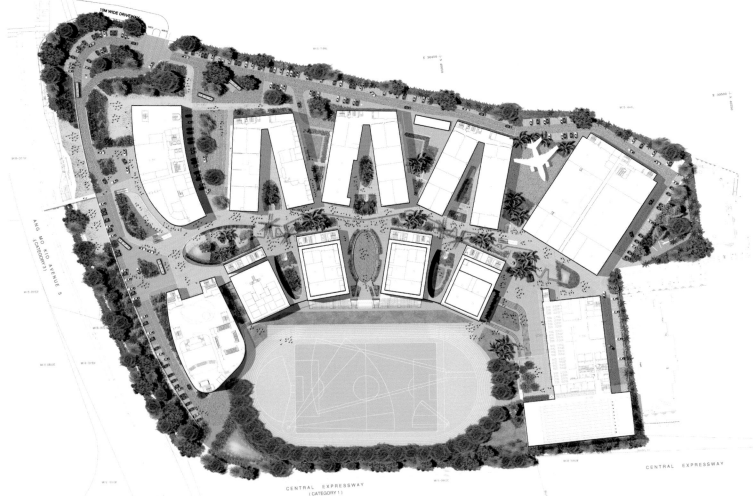

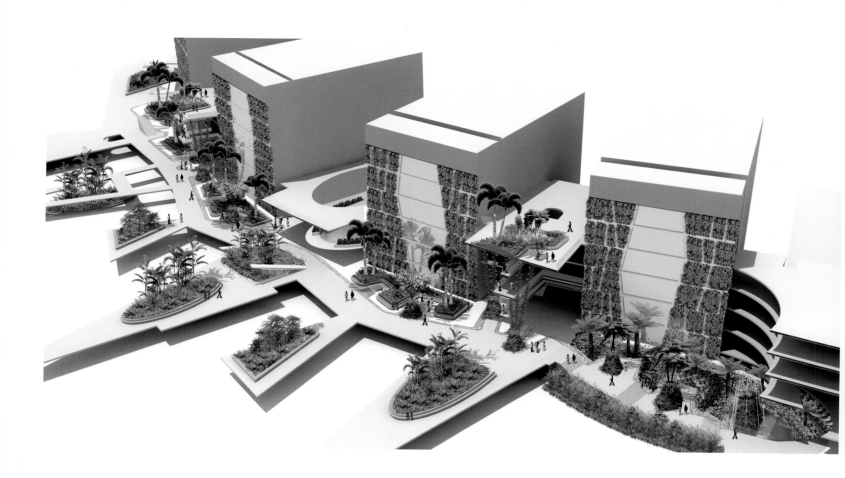
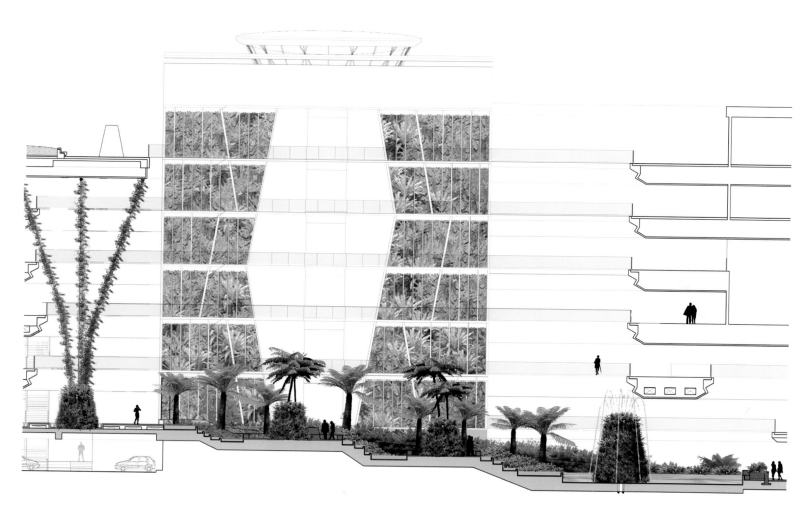

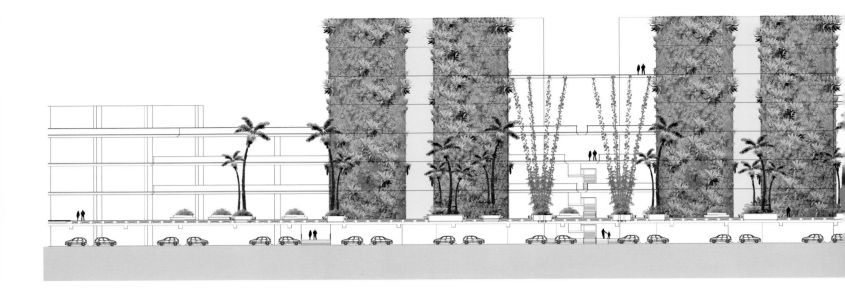

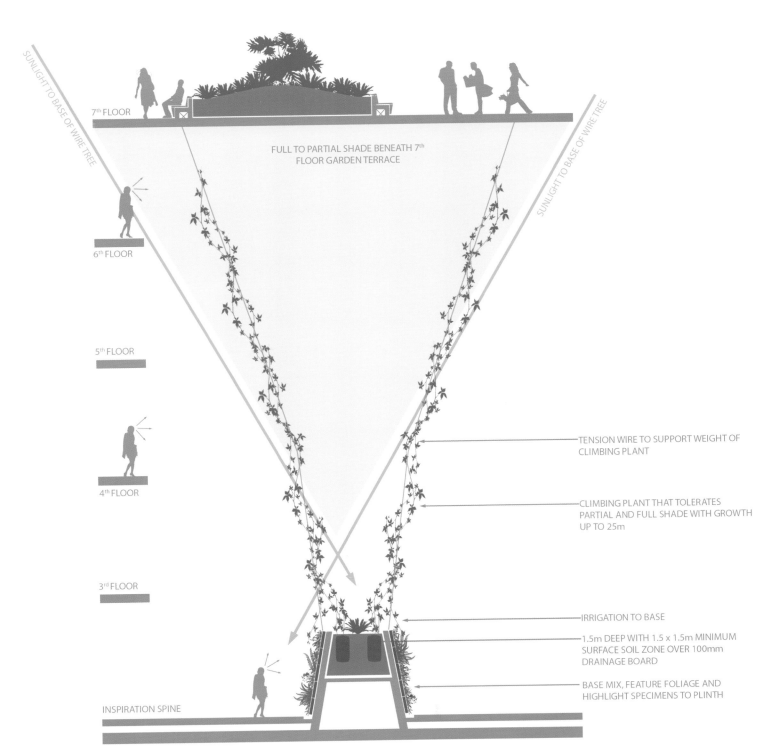

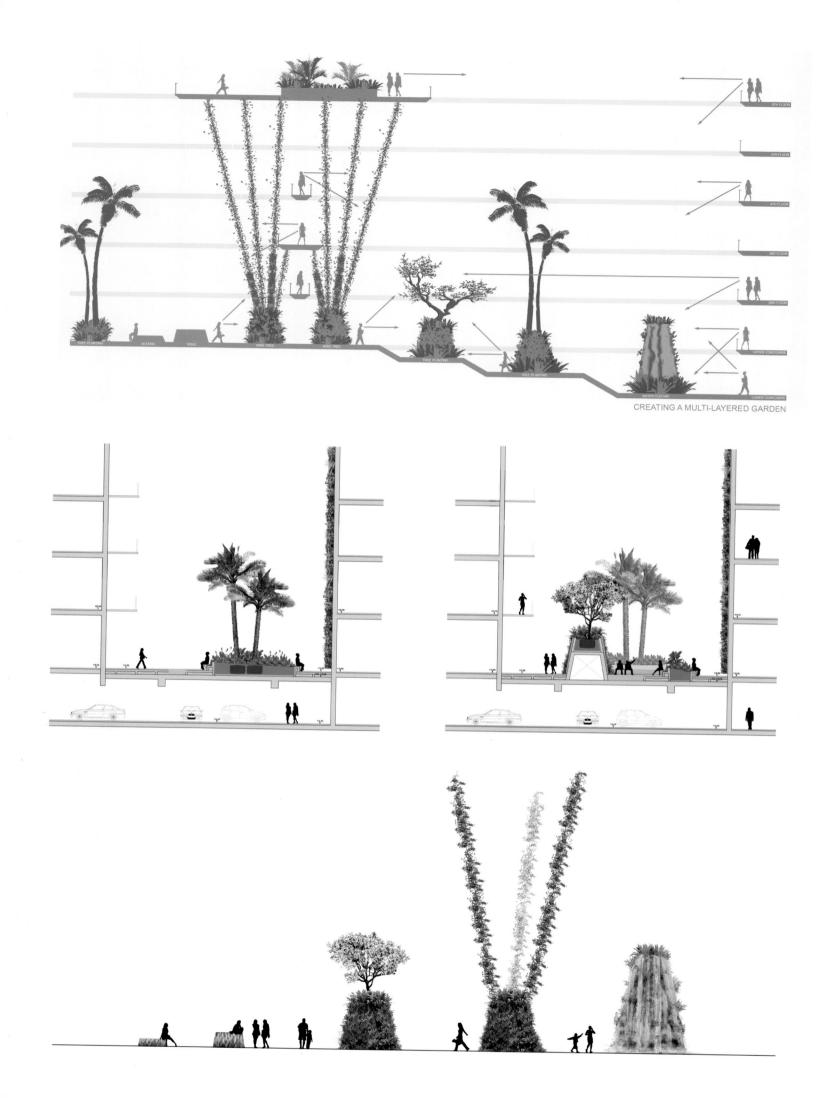

CREATING A MULTI-LAYERED GARDEN

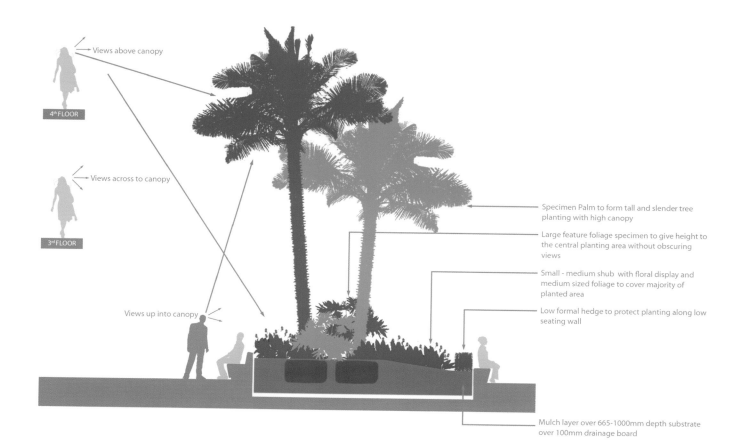

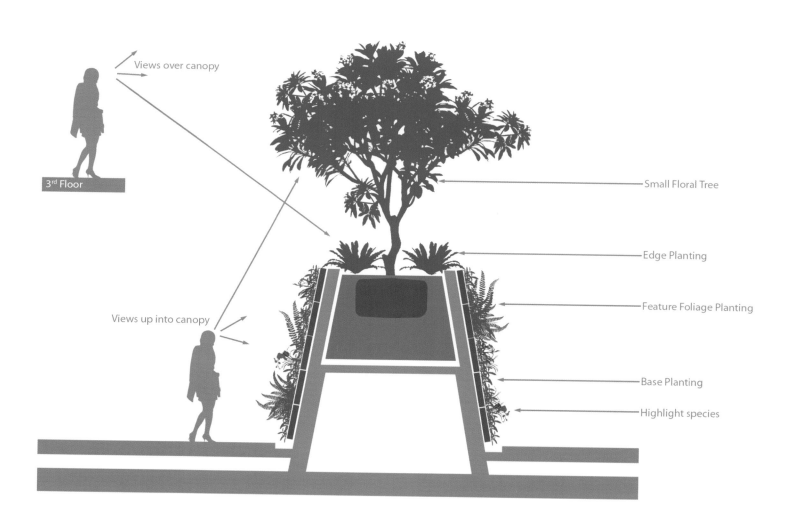

Grant Associates 事务所是中标团队的成员,他们与 RSP 一起赢得了本案,他们将设计和实施新加坡工艺教育学院中区学院和新总部的建设项目。Grant Associates 事务所承接从概念设计服务到详细的生产信息的全部工作。

中区学院是第三所,也是最后一所工艺教育学院,是一所创造和创新型的大学。设计中,建筑和景观被紧密整合在一起,并通过多个层次来确保校园空间的使用效率。一层是入口、停车场和服务区;二层即人们熟知的"枢纽",提供了导航服务和社交空间,并通过几条空中走廊和平台连接了 12 个教育专区。

"枢纽"中包括一系列花园和基座,包括了标本树木和五颜六色的"线树"。这些花园和基座围绕垂直堆叠的空中走廊创建了独特的空间和关键的循环节点。水从花园的源头一直流到终点花园,为花园提供了阴凉的休息空间。一系列平台和下沉的花园与"枢纽"相连,为学校的员工和学生们提供了许多空间,通过这些空间人们可以通往更广阔的校园景观。

中区学院的校园景观为 10 400 名学生和 850 名职工而设计。新工艺教育学院总部设计也与校园设计相映成趣,其配套会议设施可同时接待 550 名员工。

Location
Singapore

School of the Arts
艺术学院

Landscape Architect
Cicada Pte. Ltd.

Building Architect
WOHA

Photographer
Patrick Bingham-Hall

This project is a hybrid between a specialist arts high school and performing arts centre, and is a machine for breezes, located in the dense, tropical inner city of Singapore. The School of the Arts, Singapore (SOTA) is thoughtfully designed to provide not only a safe and stimulating environment for learning, but also places of delight for the public.

The podium contains a music auditorium, a drama theatre, a black box theatre and several informal performing spaces. To enhance the vibrancy of the city, shops are provided along the external covered walkway and a large civic amphitheatre is created under the canopy of large conserved trees. The sectional relationship between gathering spaces on different levels allows for easy ventilation and a comfortable microclimate, with barrier free access incorporated throughout the building.

The academic blocks are designed for natural ventilation with breezeways in-between the blocks. Gardens on the top of decks cut out heat gain, absorb carbon, and provide shady outdoor break-out spaces and play areas, while green facades cut out glare and dust, keep classrooms cool and dampen traffic noise. These seamless indoor-outdoor spaces with comfortable microclimates allow different sized groups to interact and relax without leaving the secure environment of the school.

　　该项目是一所专业艺术高中和表演艺术中心的集合。它位于人口稠密、气候酷热的新加坡市中心，犹如一个能产生凉风的机器一样。新加坡艺术学校（SOTA）精心的设计不仅为学习者提供了安全和激励的学习环境，同时也是公众开心游玩的地方。

　　裙楼包括一个音乐厅、一个戏剧剧场、一个黑匣子剧场，还有一些小型非正式表演空间。为加强城市的活力，建筑物外部沿着人行横道布满商店。在古树林荫的庇护下，形成了城市的圆形剧场。不同层面的聚集场所被分成各部分区域，各个部分都毫无障碍地与整个建筑物相联系，从而保证各部分空气流通便利，局部气候宜人。

　　学术区为自然通风，在各区之间建有通风道。平台上面的花园可以切断热量吸入，吸收碳元素，并且可以提供室外放松或玩耍的阴凉地。同时，绿色的外墙可以阻断强光和灰尘，保持室内凉爽并抑制交通噪音。这些密封的室内外空间，小气候舒适宜人，使许多不同大小的群组不用离开学校安全的环境，就能进行交流，放松自己。

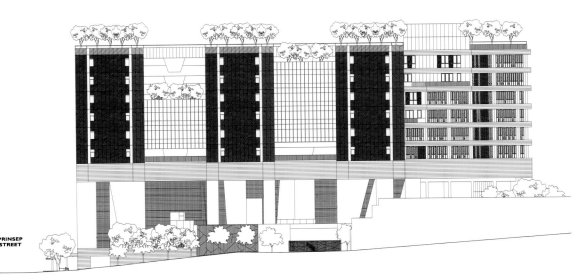

PRINSEP STREET

EAST ELEVATION
0 10 20 50 M
1 : 600

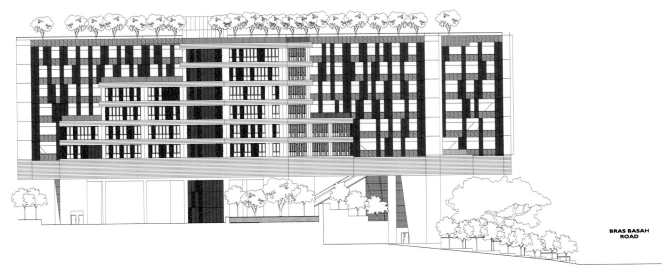

BRAS BASAH ROAD

NORTH ELEVATION
0 10 20 50 M
1 : 600

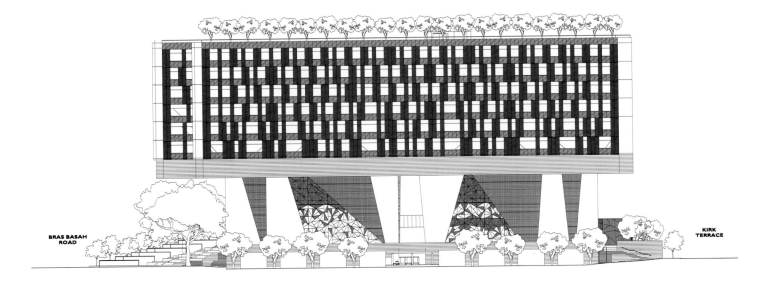

BRAS BASAH ROAD

KIRK TERRACE

SOUTH ELEVATION
0 10 20 50 M
1 : 600

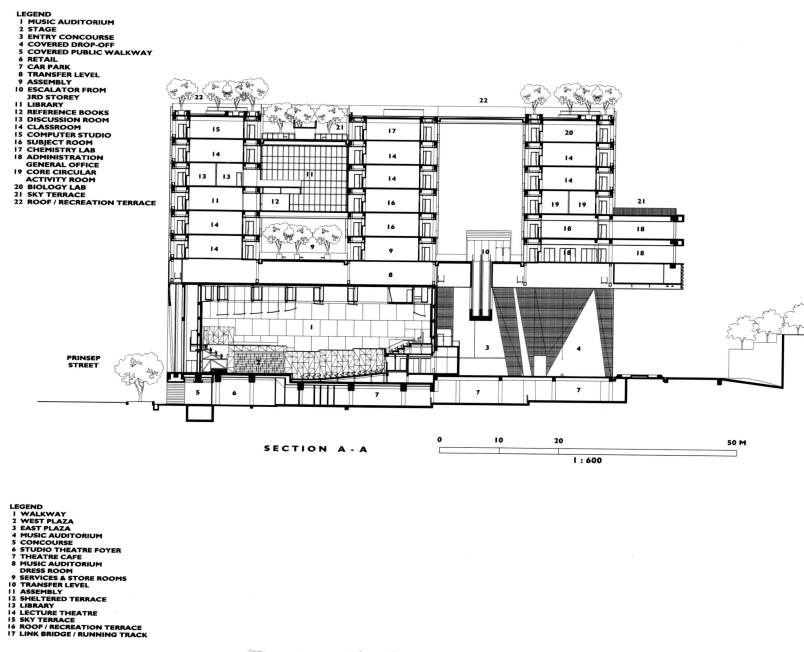
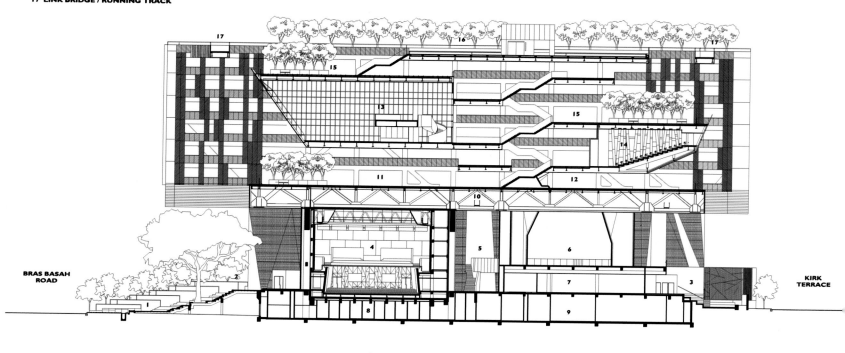

Location
Singapore

Jardin
Jardin 项目

Building Architect
DP Architects

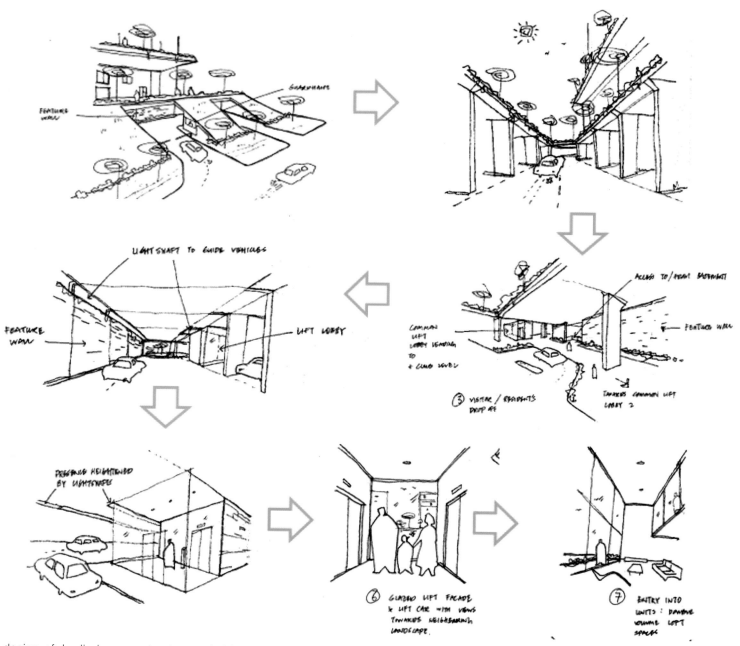

The design of Jardin incorporates two main ideas – a "vertical" garden setting and the French notion of living. Extending the desire of enjoying a garden setting in front of one's abode, Jardin allows this simple pleasure of living in a garden to be materialised in a high-rise environment, by literally bringing gardens right up in the sky.

At each alternate level, extensive gardens extend from the loft units, serving as deep communal "balconies". Besides functionally providing shade and buffer from city noise, these gardens connect the units' living spaces, allowing for use as social spaces in the open air. The roof hosts a club and recreation garden with a collection of event pavilions that also intend to cultivate gatherings.

To compensate for a ten-storey height constraint, the Jardin's design utilises large floor plates and a lengthy building perimeter in an exploration that merges communal landscape with high-density urban living.

By negotiating planning code guidelines and the relationships between softscape and hardscape, extensive facade-length terraces have been built into every alternating level to create double-height garden spaces accessible to each of the building's residences. A number of techniques in landscape design are explored to provide a blend of experiences by which the garden evolves about a user; these implementations of natural spatial development have the potential to establish new forms of high-density urban living in the tropics.

Such practices of garden design include serial vision, developed by architect and urban designer Gordon Cullen in the English Townscape movement, which considers the garden as a sequence of spaces that reveal themselves to a meanderer in succession: in the Jardin, garden elements — green walls or feature walls— serve as focal points for the reorientation of the visitor along various axes of travel; shafts of natural sunlight are employed to vary visibility. The complexities of the garden channel a sensory journey.

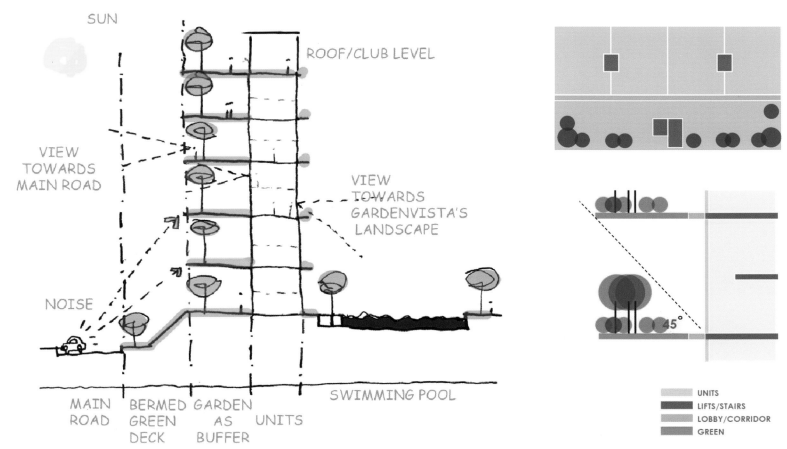

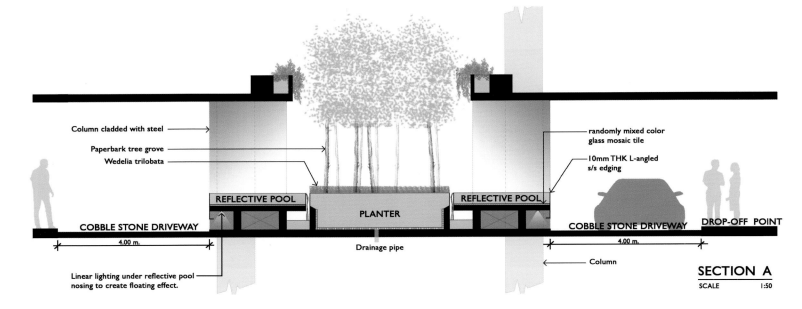

FRONT PERSPECTIVE

WESTPERSPECTIVE

Jardin 项目的设计体现了两个概念——一座垂直的花园和一种法式生活理念。Jardin 项目将花园直接建到空中，使人们得以在高层建筑中享受生活在花园里的乐趣，在住所前尽享花园景致。

本案中，花园作为公共的"阳台"，每隔一层从阁楼单元里延伸出来。花园除了可以遮阳，还可以隔离城市噪音，将单元生活空间连接起来，使人们可以在露天使用公共空间。屋顶有一间俱乐部和休闲花园，花园里有许多亭子，可以举办聚会。

由于受到 10 层的楼高约束，Jardin 项目设计中使用了大块楼面板和较长的建筑总长，旨在探索如何整合公共景观和高密度的城市生活。

本案设计力图在软景观和硬景观之间达成和谐，在建筑表面，每隔一层都建有一个交替平台，来创造双倍高度的花园空间，楼里的每户居民都可以进入这些花园空间。本案同时还探讨了一系列的景观设计技术，为如何围绕客户来发展公园提供多方面经验。通过探索利用自然空间，本案有可能为人们在热带地区高密度的生活提供新模式。

本案的花园设计体现了建筑师和城市设计师 Gordon Cullen 的愿景，他曾参与英国市容运动，该运动认为花园是空间的一个序列，为畅游其中的人陆续展示自己。在 Jardin 项目中，花园元素——绿墙或特色墙——为沿着不同路径观光的游客提供重新定位的焦点，不同强度的自然光可以调节可见度。花园的复杂通道为游客提供了丰富的感观享受。

Location
Bangkok, Thailand

Hansar

汉莎项目

Landscape Architect	Building Architect	Local Architect	Photographer
WOHA	WOHA	Tandem Architects	Patrick Bingham-Hall

Hansar Bangkok is located in an upmarket area in the heart of Bangkok. The site is surrounded by luxury hotels, shopping areas, City park and Turf club.

The site is small and irregular, with a plot ratio of 1:10. The design captures the value of the site by maximizing the building area with a 43 storeys tower which looks out over the surrounding buildings. In Bangkok, living up high provides less noise, less dust, cooler breezes and privacy with full exposure to the views. The views are east towards Lumpini Park and west towards the Turf club.

To negotiate between the desire for views and the need to provide shading, a metal mesh screen has been developed. These sun screens which are made of expanded mesh also serve as privacy screen for the units. The expanded mesh, which forms the building's outer skin, is coated in metallic bronze color to create a contrast to the grey Bangkok sky.

The small site limits the potential green areas on the ground level so sky gardens have been created, equal to 30% of the site area. These sky gardens are placed at every 5 floors with staggered sky pavilions. Each unit either has a private lift lobby with an entry to the sky garden or a living room with sky garden views. Some have a private sky pavilion.

At the podium, the ground level consists of glassy retail below 6 storeys of carparking. Aboveground parking is required in Bangkok to avoid the frequent floods. Green creeper screens wrap around the carparking podium, on top of which sits the cantilevered swimming pool. Landscaping is incorporated in all the common areas throughout the building. Lower level units have private cantilevered gardens. These gardens create a rhythm of green elements which run vertically throughout the whole development.

The combination of these elements creates a unique character for the development. The golden mesh and floating greenery allude to the glimpses of gilded temple and luxuriant gardens that hide amongst the concrete jungle of Bangkok.

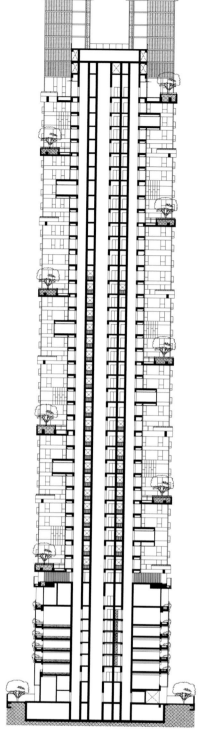

HANSAR BANGKOK
CROSS ELEVATION

HANSAR BANGKOK
GROUND FLOOR PLAN

1 : 300
W O H A
COPYRIGHT SEPTEMBER 2011

LEGEND
1 ARRIVAL LOBBY
2 DROP OFF AREA
3 RESIDENCE LOBBY
4 RESIDENT LIFT LOBBY
5 MAIL BOX
6 WALK WAY
7 REFLECTION POOL
8 HOTEL LOBBY
9 BELL COUNTER
10 BAR
11 RESTAURANT
12 OPEN KITCHEN
13 KITCHEN
14 PRIVATE DINING ROOM
15 STAIR HALL
16 OUTDOOR TERRACE
17 OUTDOOR DINING TERRACE
18 MAIN LANDSCAPE
19 DEVELOPMENT SIGNAGE
20 PARKING AREA
21 M&E ROOM
22 COFFEE SHOP

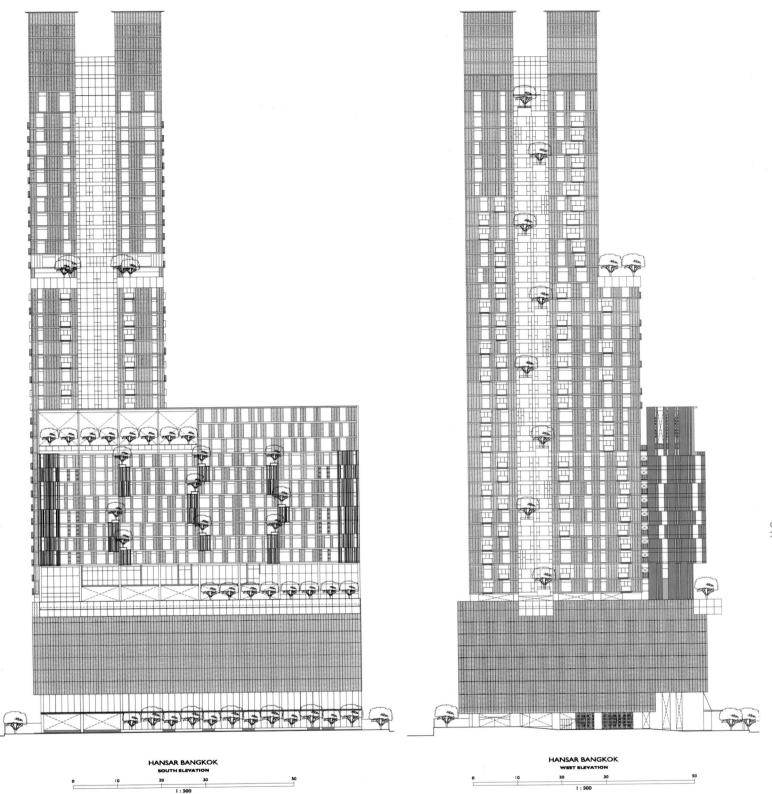

汉莎项目位于曼谷市中心的一个高档社区。周围豪华酒店林立，并配套有购物中心、城市公园和赛马场。

本案地块面积小而且形状不规则，容积率为1:10。本案旨在设计一幢43层的高楼，这样可以使建筑面积最大化。高楼俯瞰周围的建筑。在曼谷，居民住高层可以享受较少噪音、较少灰尘、凉爽微风，并视野开阔。本案东向是Lumpini公园，西边是赛马场。

该项目使用了金属网格的屏幕来遮阳，同时为大楼居民提供开阔的视野。遮阳板由扩展的网格构成，同时也为每个单元提供了隐藏的空间。这些涂成青铜色的扩展网格形成大楼的外层皮肤，与曼谷的灰色天空形成鲜明对比。

由于地面面积有限，所以设计师创建了空中花园。空中花园相当于地面面积的30%。大楼每隔五层就会设置空中花园，空中花园内还设计了错落有致的凉亭。每个单元或者有私人电梯间可以通到空中花园，或者可以从卧室看到空中花园。有些单元有私人空中亭苑。

在平台上，一楼是光亮透明的零售区，其上是6层的停车场。曼谷要求停车场高于地面，以避免被频繁的洪水淹没。绿色葡萄植物环绕着停车场的平台，平台上方是悬臂式的游泳池。景观包括了建筑物内的所有公共空间。低层单元有私人的空中花园。这些花园郁郁葱葱，贯穿了整个大楼。

本案将这些元素结合，使这个项目独具特色。金色的网格和浮动的绿色让人们联想起镀金的庙宇和华丽的花园，所有的一切皆隐藏在曼谷的钢筋水泥丛林里。

Location
Singapore

48 North Canal Road

北运河路 48 号办公楼

Landscape Architect	Building Architect	Photographer
Greenology	WOHA	Patrick Bingham-Hall

The project brief called for a new boutique office and the reconstruction of a pair of heritage-listed shophouses. WOHA was commissioned only after their demolition to reconstruct the shopfront (up to 7.5m depth) in accordance with Singapore's Urban Redevelopment Authority's conservation and planning guidelines, and to design an entirely new, contemporary rear wing.

As the original floor levels with their low ceiling heights were retained, the front end of the shophouses was deemed more suitable for meeting rooms, while the service end accommodated a mechanised carpark. The idea was to strategically lift up the open plan offices within the upper 4 floors where the floor plate size is maximised, higher headroom is gained, better views are enjoyed and more natural daylight is accessed from the sides. Every flat roof area is also transformed into roof gardens with the attic featuring the office's recreational lounge from which unblocked panoramic views of Hong Lim Park and PARKROYAL on Pickering Hotel can be enjoyed.

Unlike a typical internalised courtyard, the main design strategy was to invert the shophouse typology by carving out valuable floor area to create an externalised, urban, public pocket park at the very heart of the office instead. A café, break-out areas and meeting rooms are organised around this park, enjoying the greenery and light that it brings to the deep plan. This public gesture further serves to reduce the intermediate scale of the 9-storey building to a more intimate, human scale at the pocket park below.

The formal architectural language of fractal, triangulated geometry originated from the need to comply with authority requirements of having splayed corners as the building is bounded by three roads. This inspired a chiselled expression that was carried through in both plan and elevation, taking the form of internal angled walls and external slanted planes, revealing a concave curtain wall like that of crystal embedded in the hollow lower strata of its atrium park space. Shading was also built into the formal language by means of an integrated sun screen within the curtain wall system and a series of perforated aluminium panels.

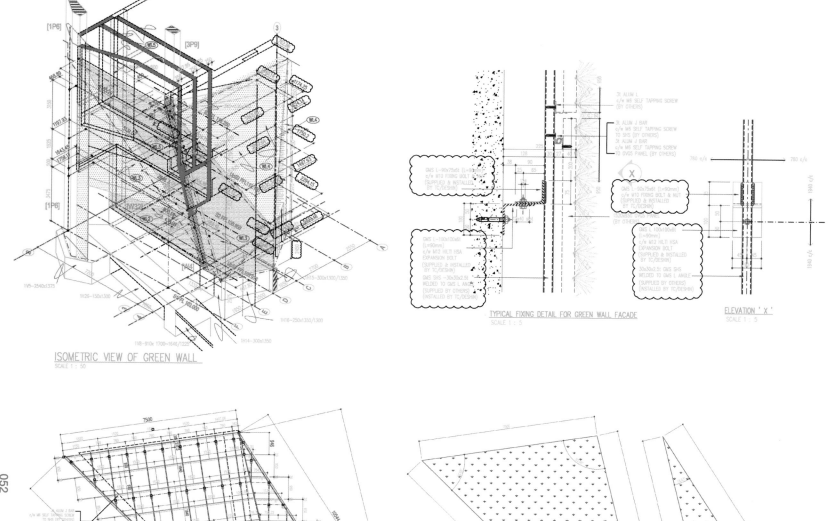

　　该项目旨在建一座新的精品办公楼和重建两家被列入遗产名录的临街店铺。旧建筑被拆除后 WOHA 接受委托，其设计任务是严格按照新加坡市区重建局重建规划指导方针重建新门面（7.5 m 进深），并为其设计一个全新的、具有同时期风格的后翼。

　　由于保留了原始楼层和较低的天花板高度，大家一致认为店铺前端更适合作为会议室，而服务区更适合建一个机械化停车场。当时的战略性想法是把办公室设置在较高的四个楼层，那里的楼板尺寸被最大化。这一设计不仅增加了室内的高度空间，而且视野较好，更多的自然光能从侧面照射进来。每一处平坦的屋顶都变成了屋顶花园，附带的阁楼也成为了一个娱乐休息室，在这里可以一览无余地欣赏到芳林公园和皮克林宾乐雅酒店的全景。

　　与典型的内化庭院不同的是，主要的设计策略是在于改变店铺的格局，通过开拓有价值的建筑面积，设计师在办公室的中央位置建立了一个小型外部公园。咖啡店、休息区和会议室都是围绕这个公园建立的，便于人们在享受咖啡的同时，享受周围的绿植和阳光。这种公开的态度将这幢 9 层高楼的中等规模缩小至更亲切、更人性化的小型公园的规模。

　　鉴于该建筑被三条街道限定了边界，重建局要求有向外伸展的拐角，因此该建筑采用了三角形几何结构这一建筑形式语言。这种结构形成了轮廓鲜明的外观，并贯穿于平面与立面。采用了内部带角度的墙壁与外部倾斜平面的形式，使凹面幕墙好像水晶镶嵌于中庭公园的中空较低层。通过幕墙系统中的一块集成日光屏和一系列多孔铝板，阴影也被设计到形式语言中。

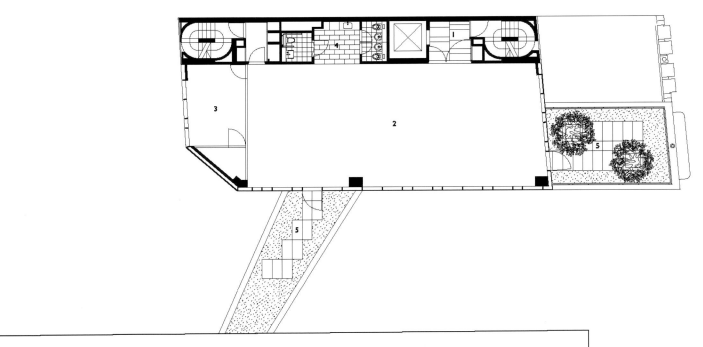

LEGEND

1. LOBBY
2. OPEN PLAN OFFICE
3. MEETING ROOM
4. PANTRY
5. ROOF GARDEN

NORTH CANAL ROAD
FOURTH FLOOR PLAN
0 1 2   5   10M
1 : 200
WOHA
COPYRIGHT APRIL 2013

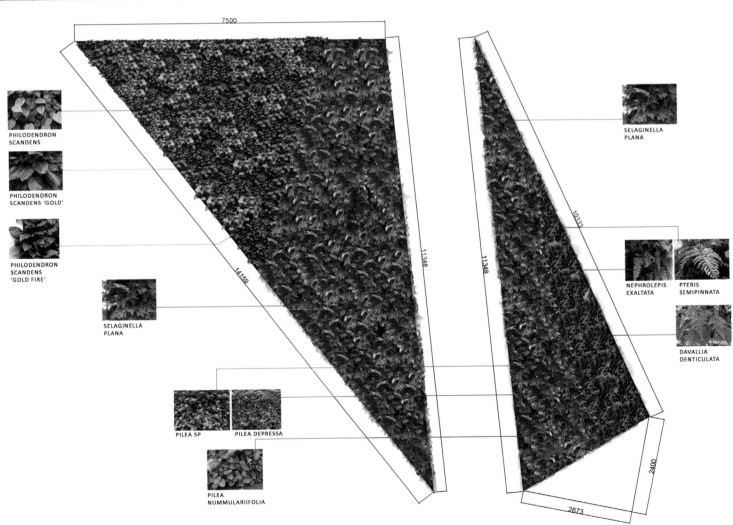

Location
Singapore

Double Bay Residences
双湾雅居绿化项目

Landscape Architect
Salad Dressing

Building Architect
DP Architects Pte Ltd

Green Wall System Manufacturer
Elmich Pte Ltd

Area (Green Wall)
196 m²

Photographer
Elmich Pte Ltd

Double Bay Residences won the Excellence Award at the Skyrise Greenery Awards 2013, organised by the National Parks Board of Singapore to honour projects which demonstrate excellence in skyrise greenery designs that integrates green elements into the built environment.

This 646-unit condominium comprises 14 blocks of residential units and a multi-storey carpark. Using an engineered modular living wall system, all five levels of the Clubhouse facade and a free-standing wall at the entrance to the condo are turned into attractive living walls.

Vertical greening 15 metres of the facade, covering 140 m² and spanning five levels, of the condominium's clubhouse has made it a talking point while a 56 m² VersiWall® GM Green Wall stands majestically at the main entrance of the condo to welcome residents and greet visitors.

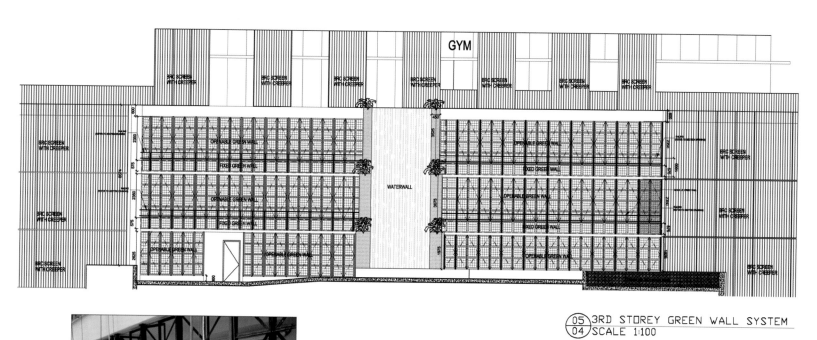

05/04 3RD STOREY GREEN WALL SYSTEM
SCALE 1:100

IMAGE REFERENCE

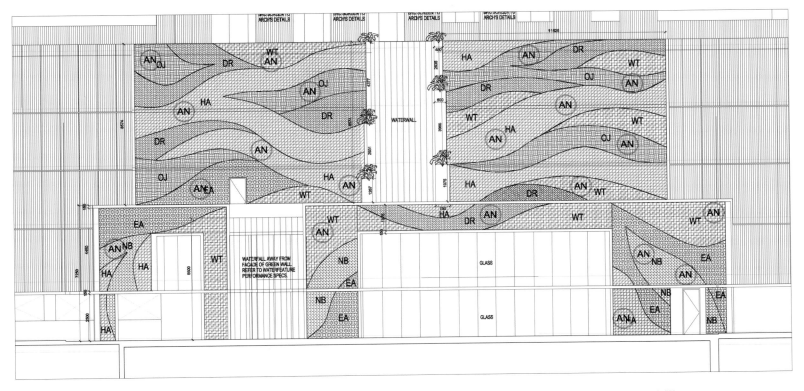
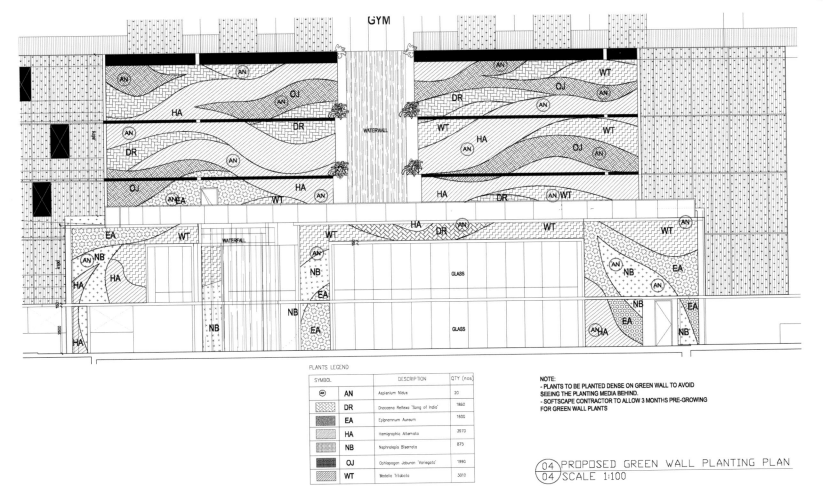

04 PROPOSED GREEN WALL PLANTING PLAN
SCALE 1:100

NOTE:
- PLANTS TO BE PLANTED DENSE ON GREEN WALL TO AVOID SEEING THE PLANTING MEDIA BEHIND.
- SOFTSCAPE CONTRACTOR TO ALLOW 3 MONTHS PRE-GROWING FOR GREEN WALL PLANTS

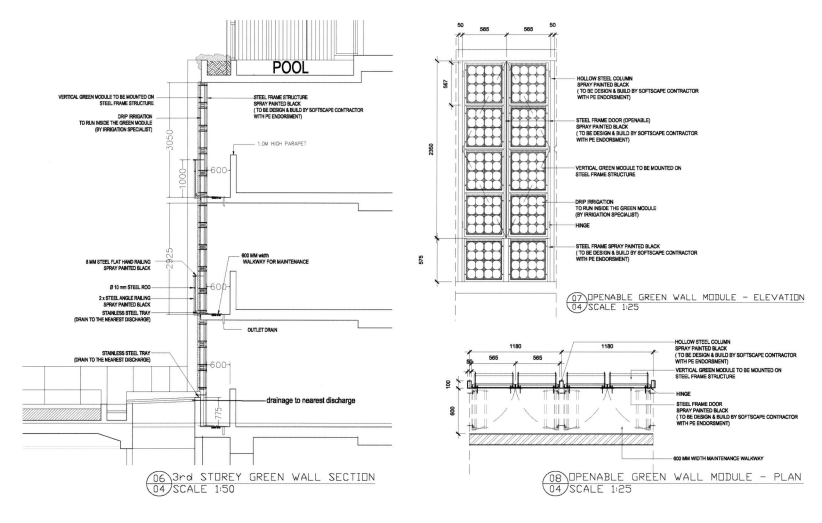

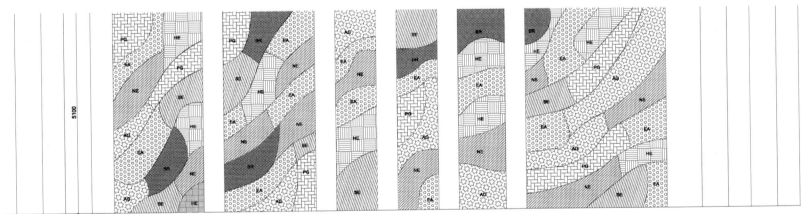

05 ELEVATION
002 SCALE 1:50

KEY LEGEND

SYMBOL		DESCRIPTION	QTY (nos)	QTY (nos) new
	AG	Aglaonema Nitidum 'Silver Queen'	88	513
	BR	Bromeliad 'Red'	93	239
	EA	Episcia Cupreata	278	709
	HE	Hemigraphis Alternata	320	397
	NE	Nephrolepis Exaltata	238	609
	PG	Philodendron 'Yellow'	274	421
	SE	Selaginella Sp	270	382

双湾雅居获得了2013年由新加坡公园局组织的空中绿意奖的优秀奖，这个奖项是为了表彰那些在建筑环境中融入绿色元素并表现卓越的空中绿化设计的。

这个拥有646个住宅单元的公寓楼包括14幢住宅户型和1个多层的停车场。5层楼的俱乐部立面和公寓入口处的一面独立墙，利用设计好的组合式活体墙系统，都变成了植物活体墙结构，非常引人注目。

公寓楼俱乐部的表面垂直绿化达15 m，覆盖140 m²，横跨5层楼墙面，这些已使此建筑成为人们竞相讨论的焦点。另外，它还有一面56 m²的蔼美生态绿墙，宏伟地耸立在公寓楼的主要入口，向入住的居民表示欢迎，向来往的参观者表示致敬。

Location
Singapore

Shaw Centre
邵氏大厦

Landscape Architect
Greenology

The 555 m² of GVG on this shopping mall in the heart of Orchard Road – Singapore's busiest shopping street, is inclusive of several alternating green walls on the balconies of the office tower and an eye-catching larger green wall that covers the porch cochere above the main entrance of the mall.

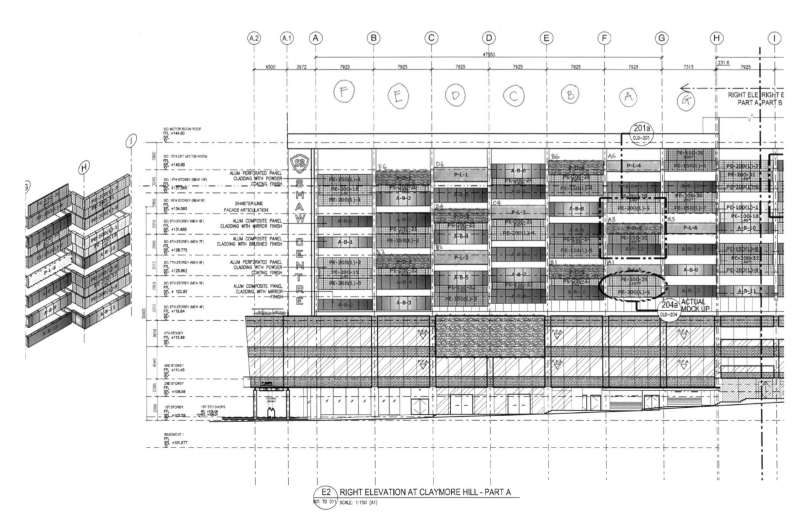

E2 RIGHT ELEVATION AT CLAYMORE HILL - PART A

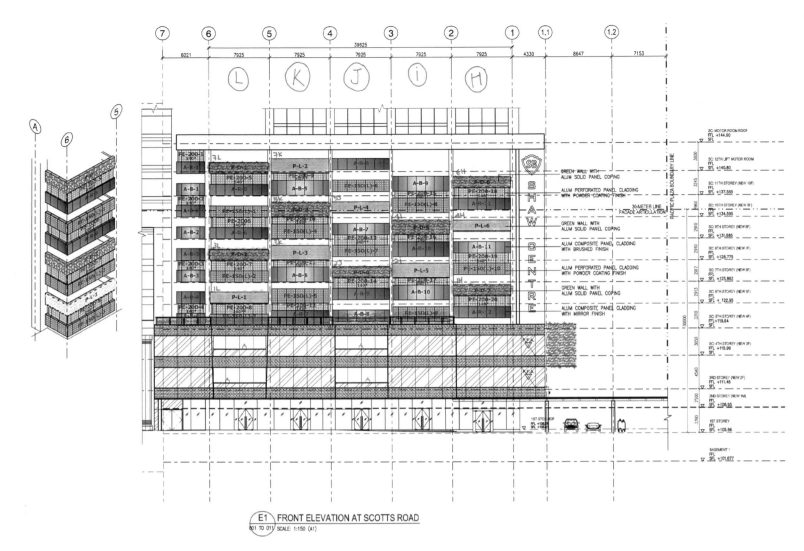

E1 FRONT ELEVATION AT SCOTTS ROAD

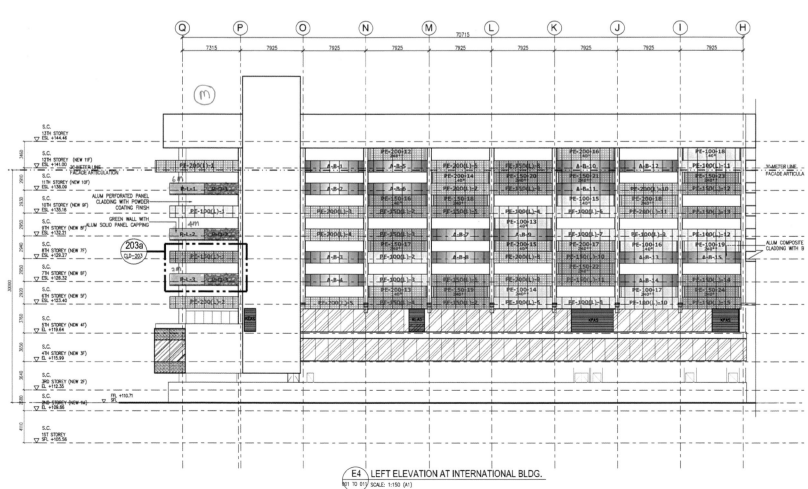

E4 LEFT ELEVATION AT INTERNATIONAL BLDG.

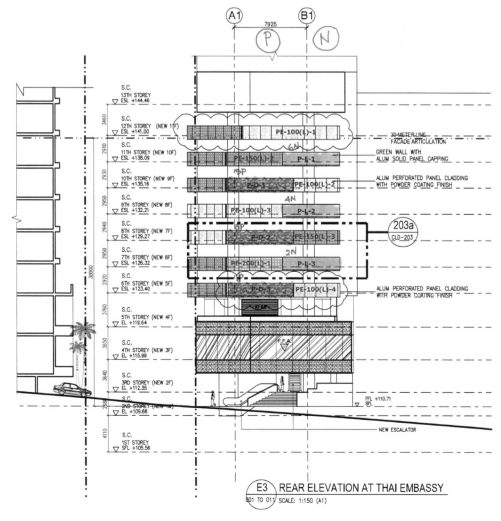

被 555 m² 的 GVG 覆盖的购物中心位于新加坡最繁忙的购物街——乌节路的中心地段。在办公大厦的阳台区域有几条交互纵横的绿墙，其中一面显眼的较大的绿墙，覆盖着购物中心正门上方的门廊。

Location
Singapore

The Heeren

麒麟大厦

Landscape Architect
Greenology

Situated along Singapore's busiest shopping street – Orchard Road, The Heeren stands out with a 6-storey-high green wall on its facade. The fire escape of this shopping mall is furnished with 435 m² of our PSB (Singapore Productivity and Standards Board) certified GVG, the first fire-rated green wall in Singapore. The lower levels of the green wall which are at eye-level with street pedestrians are filled with plant species with contrasting colours to create interesting designs.

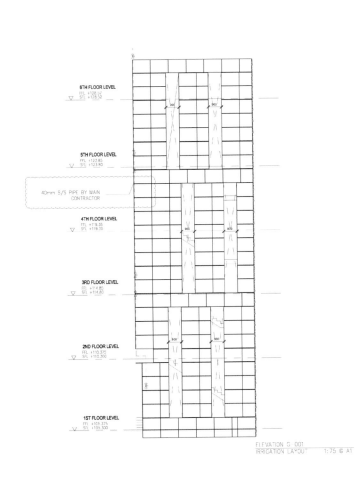
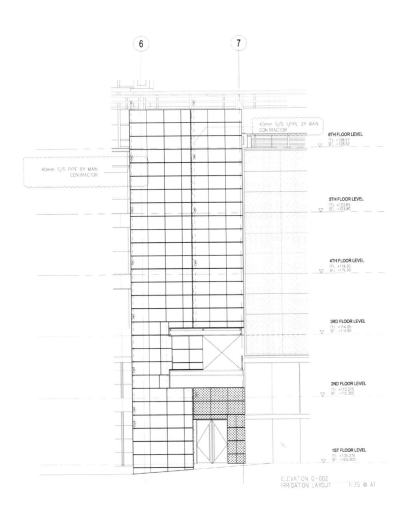

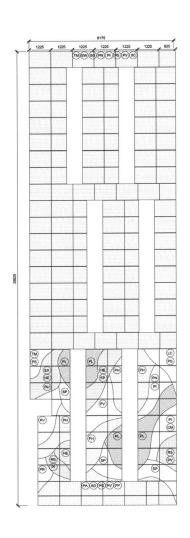
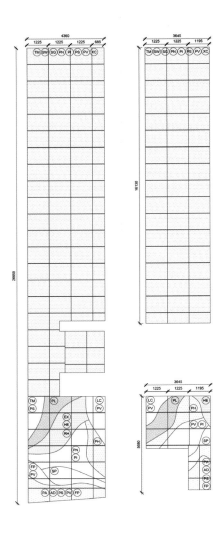
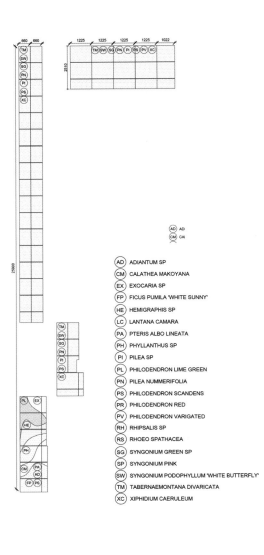

麒麟大厦坐落在新加坡最繁忙的商业街——乌节路上，其以表面六层楼高的绿墙结构而引人注目。这个购物中心的火灾安全出口是用 435 m² 符合新加坡生产力与标准局 (PSB) 认证的 GVG 所装饰，这是新加坡第一个防火绿墙。绿墙的较低部位——与街上行人视线持平的位置，填充着各种各样的植物，色彩对比鲜明，形成有趣的设计。

Adiantum sp.

Calathea makoyana

Exocaria sp.

Ficus pumila 'white sunny'

Hemigraphis sp.

Lantana camara

Pilea sp.

Pilea nummerifolia

Pteris albo lineata

Rhipsalis ramulosa

Rhoeo spathacea

Syngonium green

Philodendron lime green

Philodendron red

Philodendron scandens

Philodendron varigated

Phyllanthus sp.

Syngonium pink

Syngonium podophyllum 'white butterfly'

Tabernaemontana divaricata

Xiphidium caeruleum

Location
Singapore

Changi Airport Terminal 3 Interior Landscape

新加坡樟宜机场三号航站楼室内景观

Landscape Architect
Tierra Design (S) Pte Ltd

Building Architect
SOM

Designed in response to a world-wide competition for Changi Airport's Terminal 3, the project was built around Singapore's commitment to be a "City within a Garden". The emphasis on conserving floor space and moving the gardens to the wall made the terminal and the airport first of its kind in the world.

The holistic design of Terminal 3 introduces lush greenery, warmth, and softness into an ultra-modern, sky-lit mega-structure of stone, glass and steel. It is a demonstration of the seamless integration of gardens into buildings, and landscape into architecture.

The five-storey-high garden of vines and epiphytes hangs down over three football fields in length. In between, water glides gracefully down 20m-high laminated wall of shredded glass and stainless steel. More than 10,000 tropical plants are showcased across the green wall tapestry. Catwalks carefully hidden away allow for easy and safe maintenance of the plants.

樟宜国际机场三号航站楼的设计理念呼应了新加坡政府建设"花园城市"的理念，在全球范围的竞争中脱颖而出。本项目旨在节约建筑面积，将花园移至墙上，使这座航站楼和该机场成为世界上独一无二的设计。

在三号航站楼的整体设计中，郁郁葱葱的绿色、温暖和柔软的元素被引入超现代的、自然采光的由石头、玻璃和钢组成的巨型建筑中。建筑展示了从花园到大楼、从景观到建筑的无缝衔接。

五层楼高的藤蔓和附生植物从上面垂下来，足足有三个足球场那么长，水从 20 m 高的碎玻璃和不锈钢复合墙上悠然滑落。绿墙上展示了10 000 多株热带植物。隐藏其间的狭小通道可供人们对植物进行维护，既方便又安全。

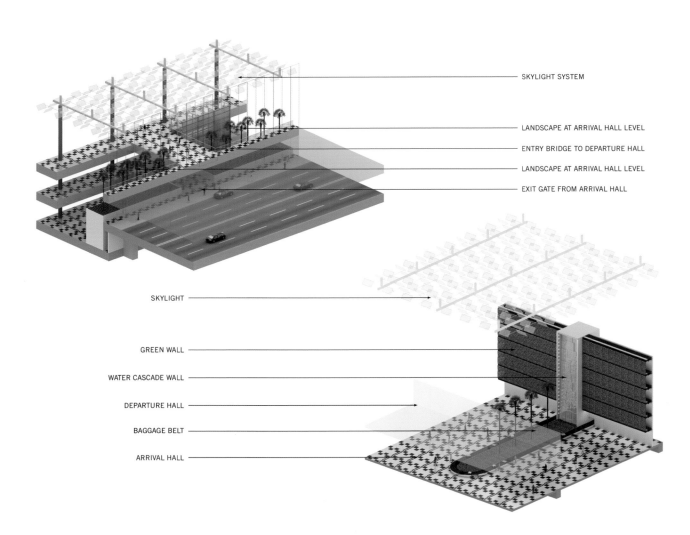

- SKYLIGHT SYSTEM
- LANDSCAPE AT ARRIVAL HALL LEVEL
- ENTRY BRIDGE TO DEPARTURE HALL
- LANDSCAPE AT ARRIVAL HALL LEVEL
- EXIT GATE FROM ARRIVAL HALL

- SKYLIGHT
- GREEN WALL
- WATER CASCADE WALL
- DEPARTURE HALL
- BAGGAGE BELT
- ARRIVAL HALL

DETAILED ISOMETRIC GREEN WALL SYSTEM

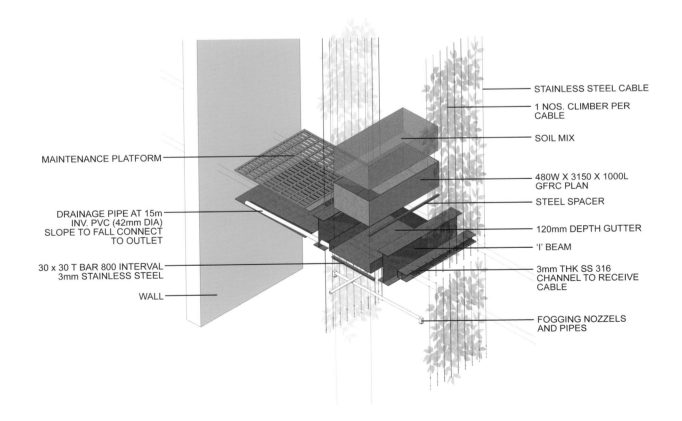

- STAINLESS STEEL CABLE
- 1 NOS. CLIMBER PER CABLE
- SOIL MIX
- 480W X 3150 X 1000L GFRC PLAN
- STEEL SPACER
- 120mm DEPTH GUTTER
- 'I' BEAM
- 3mm THK SS 316 CHANNEL TO RECEIVE CABLE
- FOGGING NOZZELS AND PIPES

- MAINTENANCE PLATFORM
- DRAINAGE PIPE AT 15m INV. PVC (42mm DIA) SLOPE TO FALL CONNECT TO OUTLET
- 30 x 30 T BAR 800 INTERVAL 3mm STAINLESS STEEL
- WALL

DETAILED SECTION OF GREEN WALL

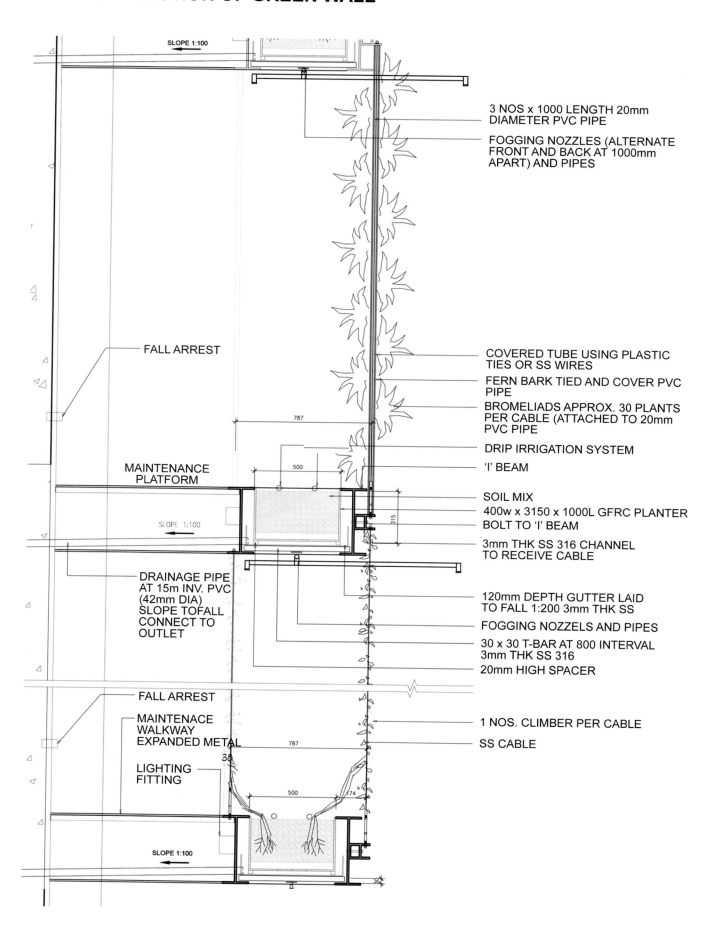

158 Cecil Street

Location
Singapore

158 丝丝街项目

Landscape Architect
Tierra Design (S) Pte Ltd

Building & Concept Architect
AgFacadesign

Before

After

Before

After

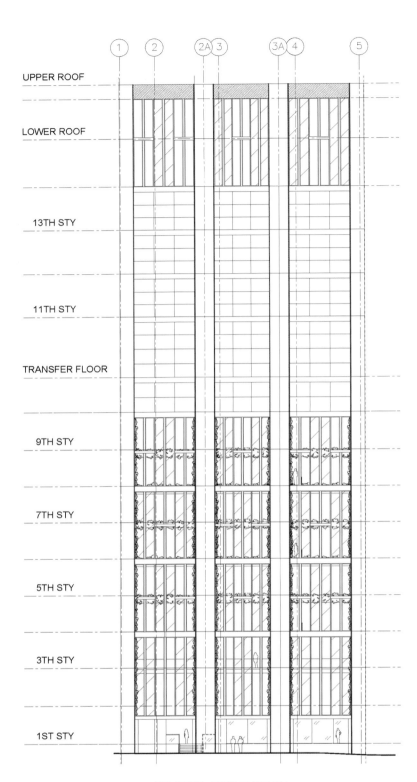

158 CECIL ST ELEVATION

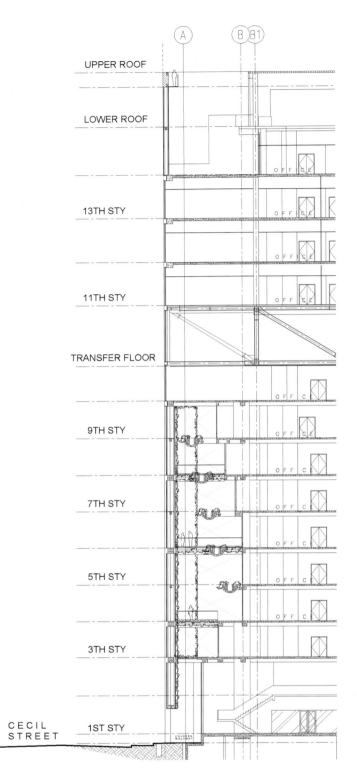

158 CECIL ST SECTION

158 Cecil Street was an existing 14-storeyed addition and alteration ("A&A") project in CBD. The task was to replace and improve an unattractive facade and multi-storeyed atrium space fronting most offices.

Designed and conceptualized by AgFacadesign Architects, the concept of a HANGING GARDEN (by day) and a GLOWING LANTERN (by night) were the architect's vision.

DAYLIGHTING AND NIGHT LIGHTING

The atrium facing east receives limited sunlight. Artificial "growth" lights are strategically mounted to simulate daylight for plant's optimum growth. The metal halide lamps of full colour spectrum are housed within highly efficient floodlights to deliver an average lighting level of 1000 lx to the plants. At planter walkways, 18W fluorescent lamps in blue and red spectrum are placed next to openable gratings.

At night, accent lights and the transparent façade visibly transform the atrium into a glowing lantern of green. Horizontal hanging plants are back-lit via LED lights reflected off the curved RC planters. The 7-storeyed high garden achieves a cathedral-like spatial quality when viewed through the glass floor at Level 3 from ground level. Façade-Architectural and crown lighting are also designed while meeting authority's night-lighting requirements.

GREEN WALL SYSTEM

Plants are chosen by the Landscape consultant (Tierra) and the contractors for their growth habit and aesthetic qualities like foliage color, leaf size, texture and shape. The light frame for potted plants houses all plumbing lines for irrigation and runoff from the drip irrigation system. Hooks at the back of each pot allow it to be mounted and removed from the grid frame for efficient maintenance. A total of 350 m² (13,000 potted plants) in the "Green Walls/Columns" is used. This is 135% of the atrium floor plate area. This is not counting another 70 m² hanging plants in horizontal planters.

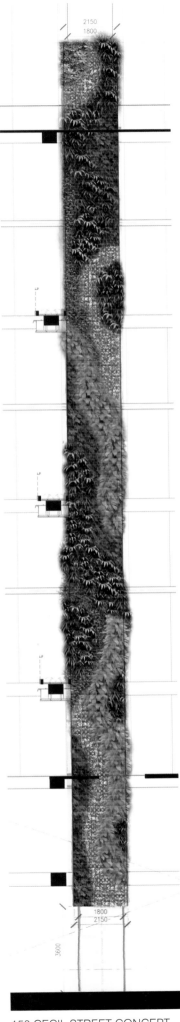

158 CECIL STREET CONCEPT

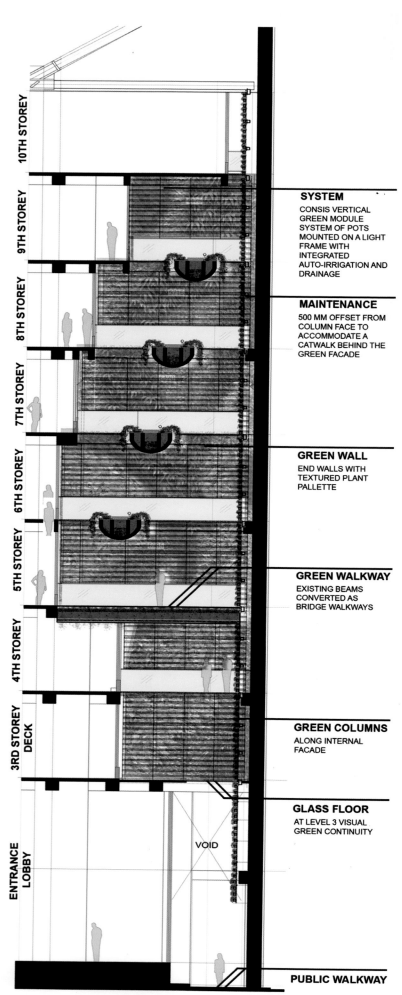

158 CECIL STREET CONCEPT

SECTION THROUGH GREEN WALL

158丝丝街项目是CBD区一座现有14层建筑的附加和变更工程。目的是取代和改进不起眼的外观和朝向众多办公室的多层门廊空间。

该项目由AgFacadesign建筑公司设计，充分体现了建筑师"空中花园"（白天）和"发光灯笼"（晚上）的设计构想。

日光和夜间照明

东向的门廊光照有限。策略性安装的人造"生长"光可以模拟阳光，为植物提供最佳生长环境。彩色全光谱金属卤化物灯被安装在高效的泛光灯内，为植物提供平均光照为1000英·小x的照明。花架走道上，挨着可开启格栅安装了18W的红蓝光谱荧光灯。

晚上，强光灯和透明的立面将门廊转换为发光的绿色灯笼。从曲面RC花架上反射的LED灯光从背后照亮了水平悬挂的植物。从一层透过三层的玻璃地板仰视，七层楼高的花园有一种大教堂的空间感。立面、建筑和树冠照明，也满足政府对夜间照明的要求。

绿墙系列

这些植物是由景观设计师Tierra和承包商根据植物的生长习惯等选择的，他们还考虑了美观因素，如叶子的颜色、叶片大小、纹理和形状。安装盆栽植物的轻型框架容纳了用于灌溉和滴灌系统的管线。每个盆栽背后都有一个钩子用来挂摘花盆，可对花盆进行有效维护。"绿墙/柱"系列共350 m²（一共13 000个花盆），达到门廊建筑面积的135%，这还不包括水平花架中的70 m²悬挂植物。

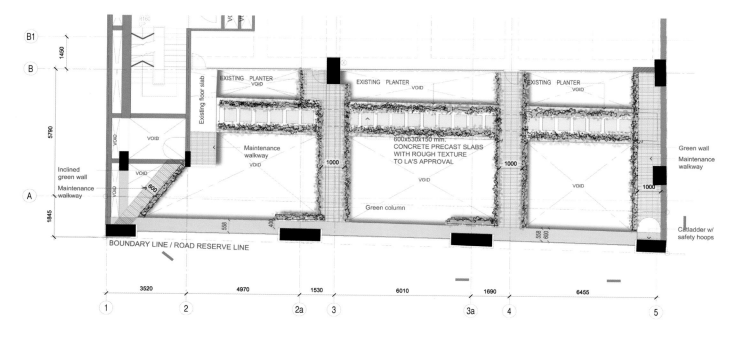

158 CECIL STREET CONCEPT

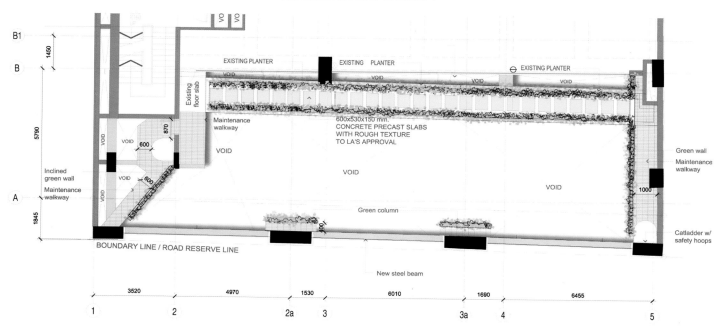

158 CECIL STREET CONCEPT

158 CECIL STREET CONCEPT

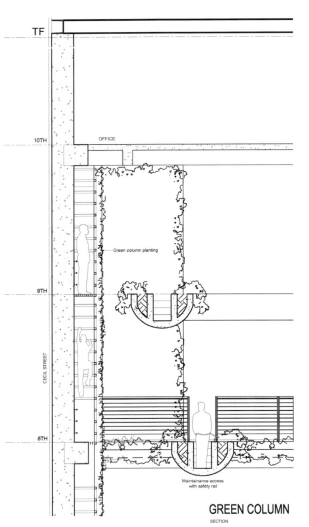

GREEN COLUMN
SECTION

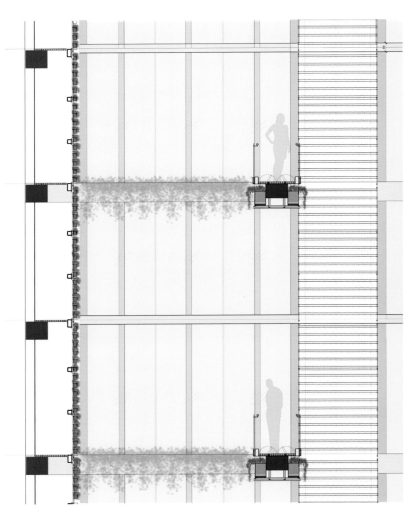

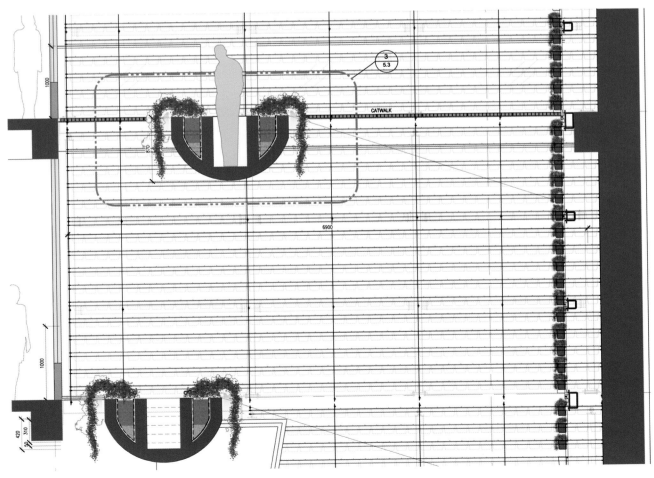

ENLARGED SECTION 2 / LH-5.3
SCALE 1:25

Location
Singapore

Jem®

Jem® 购物中心

Landscape Architect
ICN Design International Pte Ltd, Hassell

Building Architect
SAA Architects

Text
Robert Such

Wrapped in an aluminium panelled skin, the S$450 million Jurong East Mall (known as Jem) stands at the heart of a new commercial and transport hub in Jurong Lake District, a growing mixed-use business and leisure district in the west of Singapore.

Together the 70-hectare Jurong Gateway and 290-hectare Lakeside make up the new Jurong Lake District. This new neighbourhood is part of Singapore's Urban Redevelopment Authority's decentralisation strategy to build commercial hubs outside the Central Business District.

Standing next to Jurong East MRT station and bus terminal, Jem is made up of a 5-storey retail podium and an 11-storey office tower, occupied by government statutory boards.

Along Jurong Gateway Road, a series of boxes protrude from the building envelope. Clad in aluminium panels, they house usable retail spaces, and are designed to create a highly articulated and visually interesting facade.

As part of a landscape replacement strategy, the office tower has a green roof and vertical mesh panels with plants growing up them on the outdoor office balconies. The landscaped podium roof at the north-east end of the building can be used as an outdoor event space. And spread across levels five, six and seven, a mix of soft and hard landscaping further reinforces Jem's deep green credentials. Jem is the first mixed development to receive the Singapore Building Construction Authority Green Mark Platinum Award.

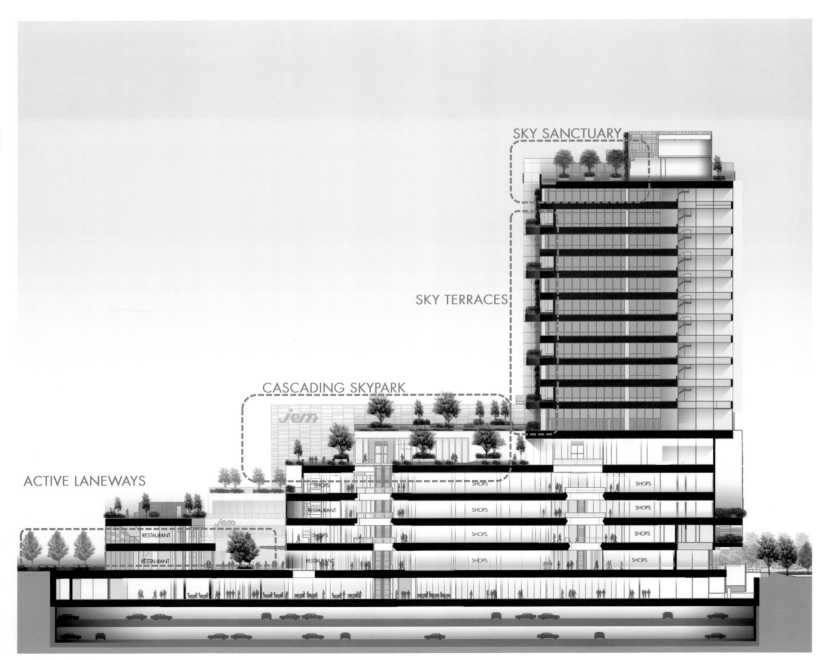

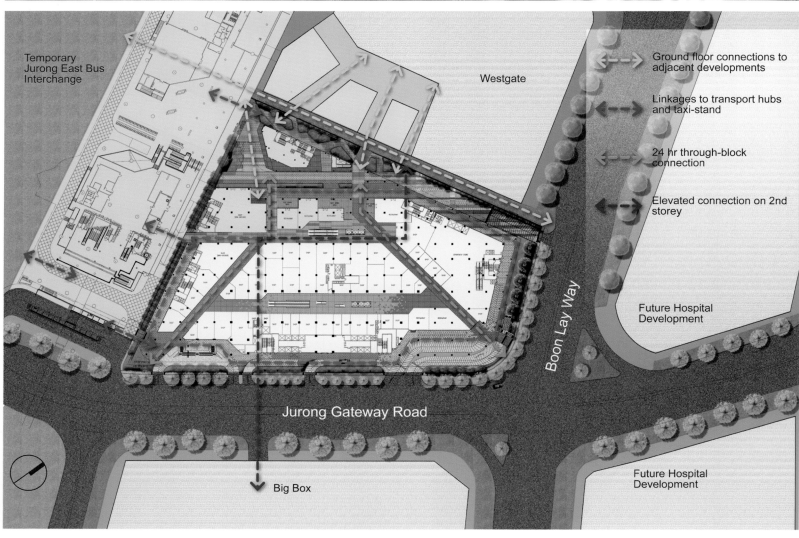

- Ground floor connections to adjacent developments
- Linkages to transport hubs and taxi-stand
- 24 hr through-block connection
- Elevated connection on 2nd storey

Temporary Jurong East Bus Interchange

Westgate

Boon Lay Way

Jurong Gateway Road

Future Hospital Development

Future Hospital Development

Big Box

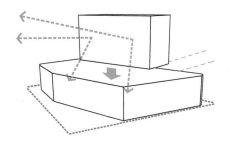 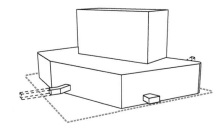 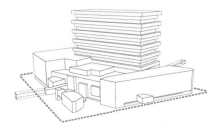

DIAGRAM

ELEVATION 1

ELEVATION 2

ELEVATION 3 WHITE POD

ELEVATION 4

Jem 项目位于裕廊湖区一个新的商业和交通枢纽中心，裕廊湖区是新加坡西部新兴的综合商业和休闲区。项目为价值 4.5 亿新元的裕廊东商城（简称 Jem），商城外面包裹的是铝制板。

新的裕廊湖区由 70 公顷的裕廊商业区和 290 公顷的湖滨区构成。这一新区是新加坡市区重建局去中心化策略的一部分，旨在建立一个中心商业区之外的商业中心。

Jem 项目靠近裕廊东地铁站和公交总站，由一座五层的零售商场和一座十一层的办公大楼构成，新加坡国家发展部在此办公。

沿着裕廊商业区路，建筑外表有一系列突出的盒子，这些盒子外镶嵌的是铝制面板。它们可以作为零售空间，当初设计这些盒子是为了创造一个高度灵活的、迷人的立面。

作为景观替换战略的一部分，办公楼有一个绿色屋顶，办公室阳台外有垂直网格板，上面种满了植物。大楼东北部景观平台屋顶可作为户外活动空间。

分布在 5、6、7 层楼的是一处软硬景观的混合体，进一步强化了 Jem 项目的环保品质。Jem 项目是第一个获得新加坡建设局绿色标志铂金奖的建筑。

Location
Kitaku, Osaka, Japan

Osaka-Marubiru
大阪 Marubiru 项目

Landscape Architect
STGK

Client
Daiwa Lease

Photographer
Takahiro Shimizu

Construction
Fujita Corporation

Area
3,250 m² (Site)
2,290 m² (Built)

Osaka-Marubiru was built 40 years ago and was the first high-rise building in the Osaka area. Beside its long history, its perfect cylinder shape is also the reason that this building has been a beloved landmark. After the building is antiquated, we are asked to renovate an open space of the building with new technology of wall greening system.

In the business district in Osaka, space for relaxation or recreation is always deficient. So is tree and plants. In this conversion project, beside increasing the coverage area of green at the site, we aim to develop the quality of greening applying a new technology of wall greening. "Canevaflor" – a new system of wall greening – has much thicker soil in the system compared to other system. It is almost vertical earth. It is the most intimate garden for people living in the city.

Regarding wall of building in the city as new earth for plants, the open space is designed under the concept of "new mother earth of the urban". The cylindrical wall is converted to green wall and edges of existing flower bed are covered by smooth corian material which is kind to human skin. Some area of the wall is coated with real earth which is called Hanchiku, a Japanese traditional wall plaster. Applying new and old technology and material into existing structure, the open space remains as a landmark in the area and appeals a vision of the relation of new city and nature.

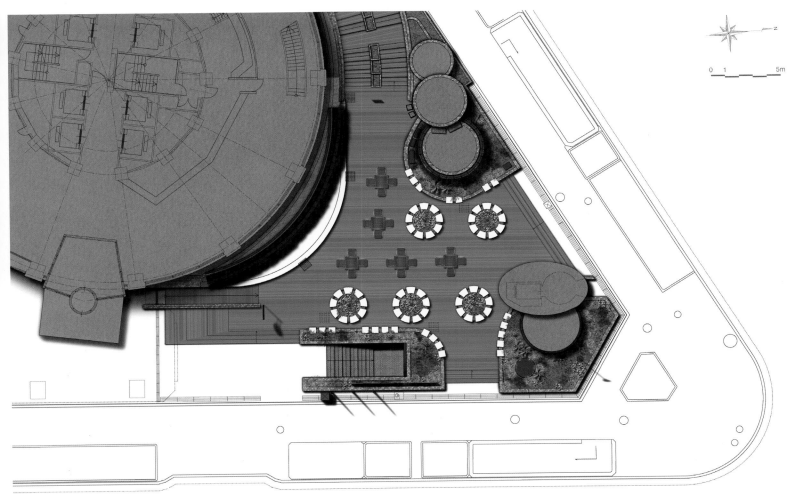

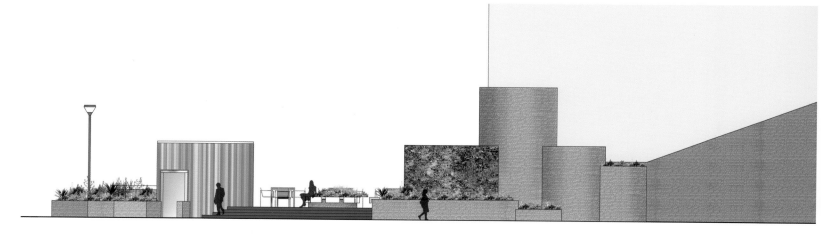

ELEVATION FROM NORTH SIDE

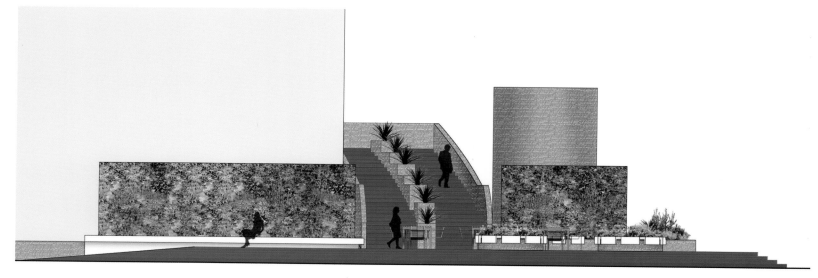

ELEVATION FROM EAST SIDE

大阪 Marubiru 项目大楼建于 40 年前，是大阪地区第一座高层建筑。因为悠久的历史，以及完美的圆柱形状，大阪 Marubiru 项目大楼成为人们喜爱的地标性建筑。由于建筑陈旧，设计师被要求使用新的绿墙技术，对大楼的开放空间进行改造。

在大阪的商业区，休息和娱乐空间相对缺乏。绿树和植物也不多见。在这个改造工程中，除了增加绿色植物的覆盖，设计师还希望通过应用新的绿墙技术——"Canevaflor"来提高绿化质量，这是一种新的绿墙技术，与其他技术相比，拥有更厚的土壤，那几乎是垂直的土地。这是生活在城市里的人们最亲密的花园。

在"城市中的新大地母亲"理念指导下，设计师设计了开放空间，将大楼的墙面作为植物生长的新土壤。圆柱墙被转换为绿墙，花圃的边缘覆盖着光滑的可丽耐材料，这是一种亲肤的材料。有的墙上覆盖了一种叫 Hanchiku 的真正土壤，这是一种传统的日本刷墙粉。本案将新老技术与材料应用于现存的建筑中，其开放空间仍旧是这个地区的地标，并且体现了新城市与自然之间的关系。

GREEN WALL SECTION

Location
Paris, France

Social Housing Croix Nivert

克罗斯克尼维特社会住房

Landscape Architect
Ciel Rouge Creation

This project of environmental housing in Paris has been ordered through an international competition in 2009 from RIVP: régie immobilière de la ville de Paris. The building has been completed in 2014.

The idea is to break from the apartment house principle and to mix them with vegetation so that the main building seems to be more like an inhabited landscape than a common building, a green landscape which gives new perspective for the future of the city. Those green facades are composed by green columns, made of earth gabions one over the other supported by a steel structure with water irrigation inside, so that real plants, even small trees can grow on those vertical surfaces. The purpose was also to complete this very mineral city part with a contemporary green architecture matching with the traditional one. As it is in the center of the city, the green purposes are very important to make the street more welcoming as there was no trees there. We can consider that this project is a landscape, urbanism and architecture project at the same time.

On the street of Croix Nivert, the main façade prolongs the old street line by a new green one giving suddenly to the city, a new image of this part a futuristic one. One part is reddish, from the brick of the old building nearby and the other part is white grey. So that the main façade is purposely a mix between those two colors. On the ground floor of this façade, in the middle, a big opening until the inner courtyard is full of sun.

In the inner courtyard the building is undulating in the southwest sunlight with different angles' views and green columns, toward the city and the subway depot with low buildings and quiet peaceful one.

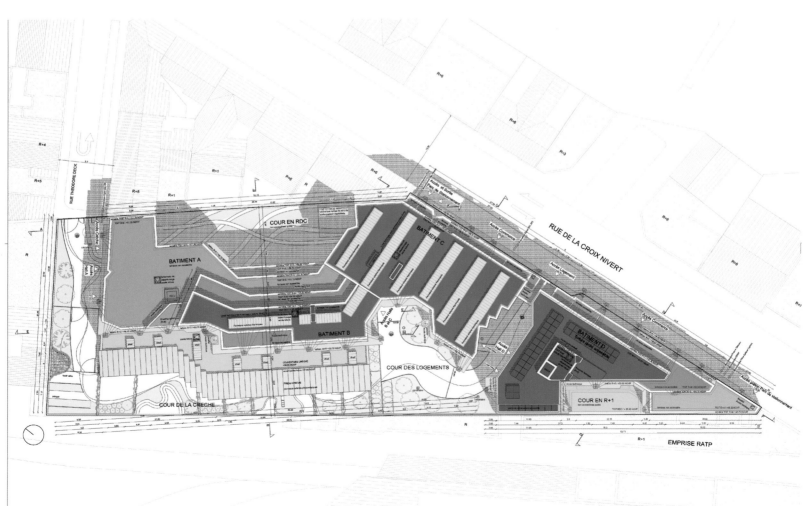

巴黎的这个环保住房项目是在 2009 年 RIVP（régie immobilière de la ville de Paris）国际比赛中被指定的，这个建筑在 2014 年已经完工。

该项目的设计理念是打破公寓住宅原则，与绿色植被融合，以使主建筑更像是宜居的风景画，而非普通的大楼，这绿色的风景赋予城市未来全新的视角。建筑绿色的外观是由绿色的柱子组成的。这些柱子由钢结构支撑的石笼组成，内部有水可以灌溉，保证植被甚至小树能在垂直面生长。该设计目的是以当代绿色建筑搭配传统建筑以完善这座矿业城市。仿佛置身市中心，绿色目的非常重要，因为它使没有树木的街道更加吸引人。我们认为这个项目是一个集都市化和建筑、景观于一体的项目。

在克罗斯克尼维特街道上，主立面通过绿色线条拉长老街道线，呈现出一片未来主义的新形象。街道一边是老建筑的砖红色，另一边是灰白色。因此主立面是这两种颜色的混合。建筑立面底层中间延伸至内院的开放空间充满阳光。

在内院，在西南方向的阳光照射下，建筑呈波浪形，能看到多个角度的景色和绿柱，面向城市和又矮又安静平和的地铁车辆段。

TERRASSE ACCESSIBLE AUX LOCATAIRES

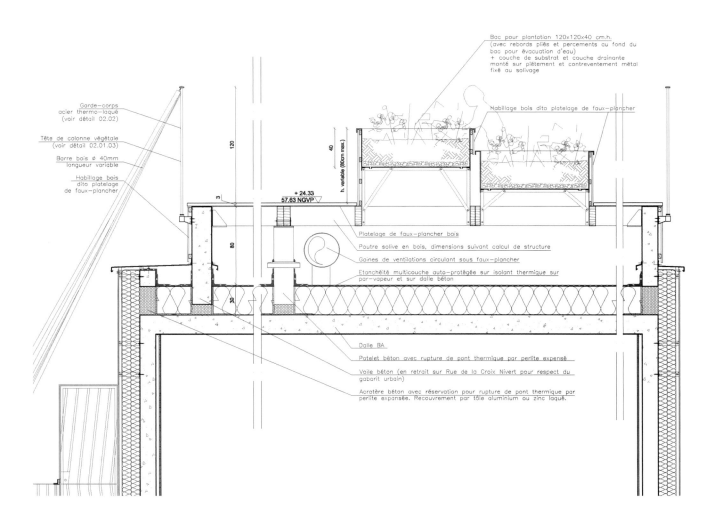

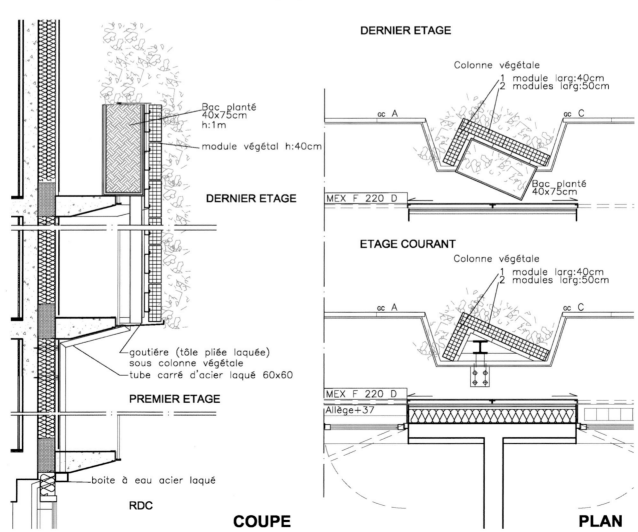

Location
Warsaw, Poland

Foundation for Polish Science Headquarters

波兰科学基金会总部项目

Landscape Architect
Adam Białobrzeski, Adam Figurski, Maria Messina

Firm
FAAB Architektura

Contractor
Doraco Construction Corporation

Vertical Green System Supplier
Semper Green Vertical Systems

Area
1,592 m² (Site), 2,180 m² (Total), 7,090 m³ (Volume), 260 m² (Vertical Green)

Photographer
Bartłomiej Senkowski

The intent of the project was to preserve the character of the 1933 dilapidated multi-family house, which had been seriously mutilated by air bombing during WWII, while converting it into an environmentally friendly and modern office space.

Simultaneously, precious architectural and historically significant elements were safeguarded alongside adhering to the restraints postulated by the Warsaw Preservation Office in regards to building mass and window layout.

The building's location, inside a sparse subdivision, established in 1930's, influenced the multilayered integration of the project within the green context and public space. The investment strove to reduce its impact on the municipal infrastructure and the natural environment.

Removing the front yard fence helped to enlarge the public space of the street and incorporate the backyard garden with it. The addition of an internal atrium and perforation through the ground level, beginning with the main entrance, established a visual link between the outside and inside, between the street and the garden.

The integrating tool to connect the building to its green context is the successful application of the first in the region external vertical garden onto both front and side elevations. The rain water collected into the retention basin irrigates the vertical garden and eliminates the demand for the municipal network. 82% of the relatively small town plot, taking into account the vertical garden, is covered with surfaces allowing natural water retention.

Access to natural light is given to 96% of the above ground level space. The ground level and underground parking lot (overhead skylights), reaches the highest possible ratio; 100% of those spaces have access to the natural light.

Preserving the original internal staircase, including the green terrazzo steps and landings filled with tiny, colorful "corset" ceramic tiles (nearing extinction) recalls the atmosphere of similar buildings of the era located in Europe.

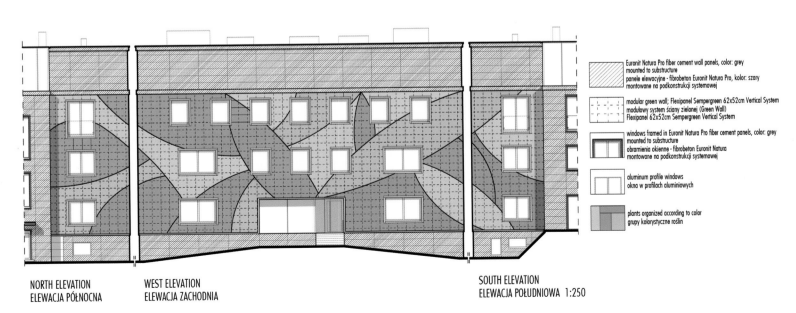

NORTH ELEVATION
ELEWACJA PÓŁNOCNA

WEST ELEVATION
ELEWACJA ZACHODNIA

SOUTH ELEVATION
ELEWACJA POŁUDNIOWA 1:250

Euronit Natura Pro fiber cement wall panels, color: grey mounted to substructure
panele elewacyjne - fibrobeton Euronit Natura Pro, kolor: szary montowane na podkonstrukcji systemowej

modular green wall; Flexipanel Sempergreen 62x52cm Vertical System
modułowy system ściany zielonej (Green Wall) Flexipanel 62x52cm Sempergreen Vertical System

windows framed in Euronit Natura Pro fiber cement panels, color: grey mounted to substructure
obramienia okienne - fibrobeton Euronit Natura montowane na podkonstrukcji systemowej

aluminum profile windows
okna w profilach aluminiowych

plants organized according to color
grupy kolorystyczne roślin

Scheme of the graphic composition of the green wall including groups of plants
Schemat kompozycji graficznej zielonej ściany z podziałem na rośliny

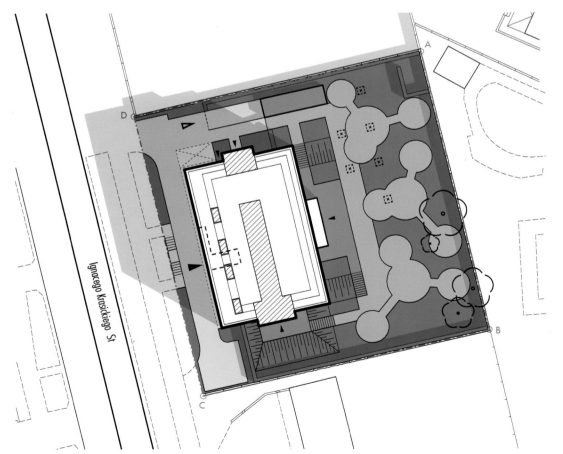

SITE PLAN / RZUT ZAGOSPODAROWANIA TERENU 1:500

 existing renovated FNP building
budynek FNP

 site boundary
granice działki

 existing surrounding buildings
istniejące budynki

 green area
teren zielony

 preserved existing trees
zachowane istniejące drzewa

 skylights lighting underground parking
świetliki oświetlenie parking podziemny

 roof skylights
świetliki na dachu

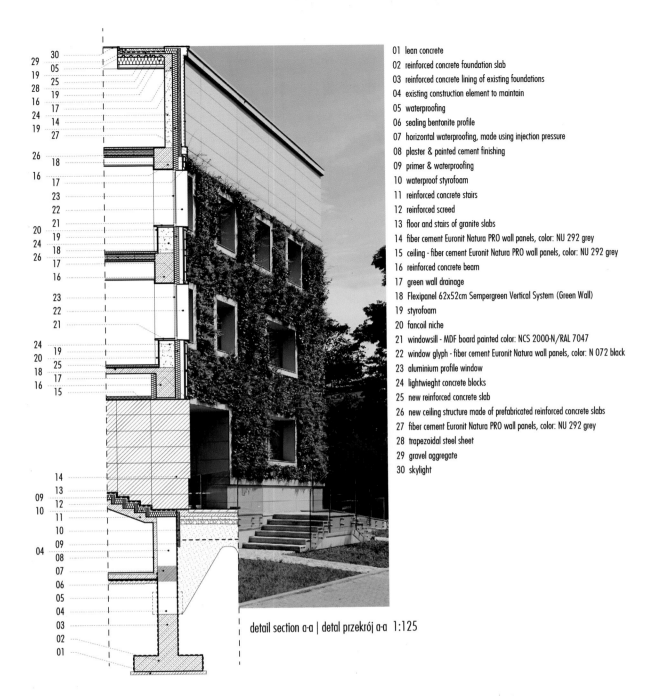

01 lean concrete
02 reinforced concrete foundation slab
03 reinforced concrete lining of existing foundations
04 existing construction element to maintain
05 waterproofing
06 sealing bentonite profile
07 horizontal waterproofing, made using injection pressure
08 plaster & painted cement finishing
09 primer & waterproofing
10 waterproof styrofoam
11 reinforced concrete stairs
12 reinforced screed
13 floor and stairs of granite slabs
14 fiber cement Euronit Natura PRO wall panels, color: NU 292 grey
15 ceiling - fiber cement Euronit Natura PRO wall panels, color: NU 292 grey
16 reinforced concrete beam
17 green wall drainage
18 Flexipanel 62x52cm Sempergreen Vertical System (Green Wall)
19 styrofoam
20 fancoil niche
21 windowsill - MDF board painted color: NCS 2000-N/RAL 7047
22 window glyph - fiber cement Euronit Natura wall panels, color: N 072 black
23 aluminium profile window
24 lightwieght concrete blocks
25 new reinforced concrete slab
26 new ceiling structure made of prefabricated reinforced concrete slabs
27 fiber cement Euronit Natura PRO wall panels, color: NU 292 grey
28 trapezoidal steel sheet
29 gravel aggregate
30 skylight

detail section a-a | detal przekrój a-a 1:125

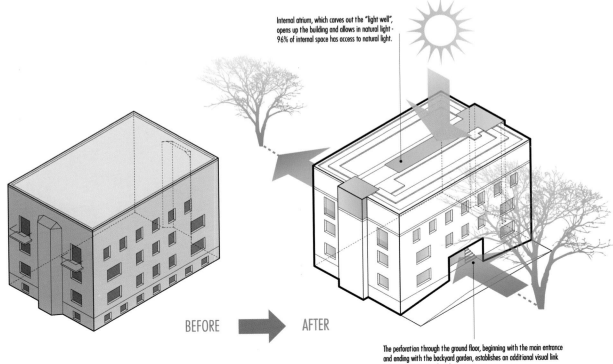

Internal atrium, which carves out the "light well", opens up the building and allows in natural light - 96% of internal space has access to natural light.

BEFORE → AFTER

The perforation through the ground floor, beginning with the main entrance and ending with the backyard garden, establishes an additional visual link between the street and garden view.

本案旨在保护这座破败的房产的特色，这套房屋建于 1933 年，由多个家庭拥有，第二次世界大战期间曾因空袭被严重损毁，本案意图将房屋改造成一个环保的现代办公空间。

同时，按照华沙保护办公室关于建筑群和窗口布局的规定，本案保护了房屋珍贵的建筑元素和历史元素。

该房屋建于 20 世纪 30 年代，位于一块人口稀疏的土地上，房屋影响了绿色环境和公共空间里的多层次整合。投资方将努力减少其对市政基础设施和自然环境的影响。

去除前院篱墙可以使大街的公共空间变大，并与后面的花园结合起来。从主入口开始，增加了一个内部中庭和穿过一层的孔墙，中庭和孔墙建立起外部与内部之间、大街与花园之间的视觉联系。

在建筑前立面和侧立面成功应用了外部垂直花园，这在本地区尚属首次，建筑与其绿色环境有机整合在一起。贮水池收集的雨水可以灌溉垂直花园，减少了建筑对市政网络的需求。82% 的小镇基地面积，包括垂直花园，都覆盖着可以收集天然水的表面。

地面上的空间 96% 都有自然光。一层和地下停车场（顶部天窗）达到最高比例，100% 的空间有自然光照。

本案保留了原来的内部楼梯，包括绿色水磨石台阶和平台，平台布满了微小的、五颜六色的像"紧身衣"一样的瓷砖（这种瓷砖在市面上已经看不到了），这样的建筑风格让人们回忆起那个年代欧洲类似建筑的氛围。

Location
Tokyo, Japan

Pasona HQ
保圣那集团总部项目

Landscape Architect
Kono Designs

Located in downtown Tokyo, Pasona HQ is a nine-story high, 215,000 square feet corporate office building for a Japanese recruitment company, Pasona Group. The project consists of a double-skin green facade, offices, an auditorium, cafeterias, a rooftop garden and most notably, urban farming facilities integrated within the building. The green space totals over 43,000 square feet with 200 species including fruits, vegetables and rice that are harvested, prepared and served at the cafeterias within the building.

The double-skin green facade features seasonal flowers and orange trees planted within the 3' deep balconies. Partially relying on natural exterior climate, these plants create a living green wall and a dynamic identity to the public. This was a significant loss to the net rentable area for a commercial office.

However, Pasona believes in the benefits of urban farm and green space to engage the public and to provide better workspace for their employees. The balconies also help shade and insulate the interiors while providing fresh air with openable windows, a practical feature not only rare for a mid rise commercial building but also helps reduce heating and cooling loads of the building during moderate climate. The entire facade is then wrapped with deep grid of fins, creating further depth, volume and orders to the organic green wall.

Pasona Urban Farm is a unique workplace environment that promotes higher work efficiency, social interaction, future sustainability and engages wider community of Tokyo by showcasing the benefits and technology of urban agriculture.

MAIN PUBLIC LOBBY

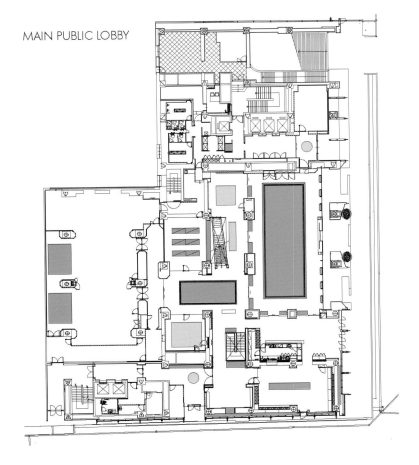

■ GREEN / FARM SPACE

■ GREEN / FARM CEILING

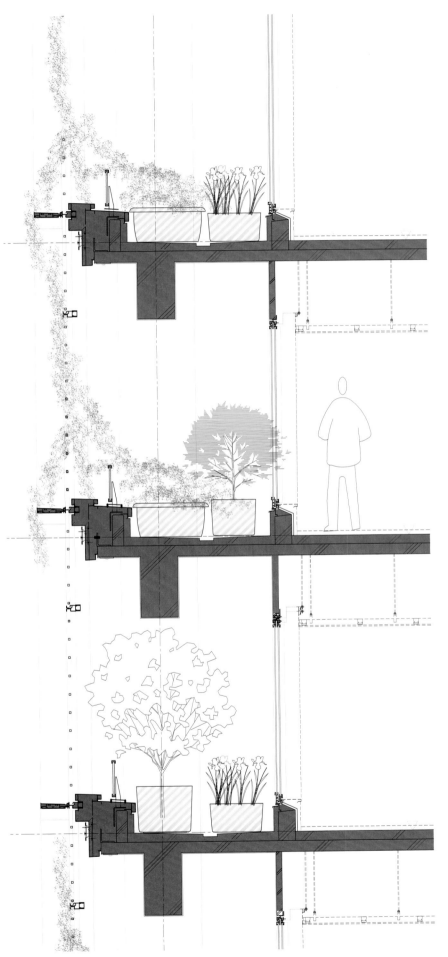

TYPICAL BALCONY SECTION

保圣那集团总部位于东京市中心，大楼九层高，面积约 2 万平方米，隶属于日本猎头公司保圣那集团。项目包括双层的绿色外墙、办公室、礼堂、自助餐厅、屋顶花园，还有大楼内集成的城市农业设施。绿色空间一共约 4000m²，种植了 200 多种水果、蔬菜和大米，可以直接被收割、加工，并在大楼内的餐厅中做成食物出售。

本案中的双层绿色外墙种植了季节性鲜花，约 1m 深的阳台上种植了橘子树。这些植物部分依靠外部自然气候，创造了一面生机勃勃的绿墙和一个动态的公共特性。对于这栋商业办公大楼来说，这一布置损失了不少可出租面积。

不过，保圣那集团坚信，公众可以从城市农场和绿色空间中获益良多，同时，这些设计可为员工提供更好的工作场所。阳台有助于内部空间的遮阳和隔热，还可以通过打开窗户来提供新鲜空气，这种实用特性在中高层商业大楼中实属罕见，在温和的气候条件下，也可以减少大楼的加热和制冷负荷。整个外墙由鳍状网格包裹，从而拓展了有机绿墙的深度、范围和秩序。

保圣那城市农场提供了独一无二的办公环境，提高工作效率，促进社会互动和未来的可持续发展，展示了城市农业的优点和技术，吸引了更广泛的东京社区。

Location
Taiwan, China

The New Construction of the Restaurant and Fitness Center in the Headquarters Building of "CSC"

中国台湾"中钢"总部大楼餐厅及健身中心新建工程

Landscape Architect
Green Empire Industrial Co., Ltd

The Inflexibility of Steel and the Harmony of Vegetation

People are always impressed by the steel industry which is characterized by being raw and cold; they seldom associate it with the natural ecology. The headquarters building of "CSC" in Kaohsiung broke up the previous thought and impression, which offered this building 8 Green Building Indexes in 2006; while the restaurant and fitness center on the other side were also with the same high-standard designs. The corporation was lucky to undertake the vertical greening of exterior wall and the design and construction of the sloping roof.

By softening the plants and integrating them with the building system, coordinating the shape of building, and making use of the color and character of the plants to highlight features of the building, the whole building maintained an advanced and natural appearance. It also promoted the overall quality of greening, reduced the indoor temperature and insulated the radiant heat. Vertical greening has become a global trend. The buildings coated with green can get positive effects both visibly (in appearance) and invisibly (in ecology).

钢的坚毅与植栽的圆融

人们对于钢铁产业的印象总是生冷的，难与自然生态联结在一起。中国台湾"中钢"位于高雄的总部大楼突破过往的思维与印象，使得这座总部大楼于2006年获得8项绿建筑指标，而其一侧的总部大楼餐厅及健身中心同样以这样的高标准规划设计。公司有幸承揽其外墙垂直绿化及斜屋顶规划设计与施工。

整体设计以植栽柔化并融合入建筑系统，配合其建筑体量造型，利用不同植栽的色彩、特性突显其体量特色，使整体外观保有前卫及自然样貌，同时提升整体绿化量及降低室内、周边温度，隔绝辐射热滞留。垂直绿化已是世界性的潮流，披上绿色外衣的建筑，在有形（外观）及无形（生态）中都能得到正面效益。

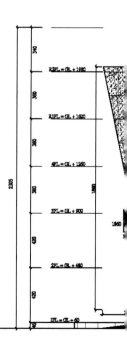

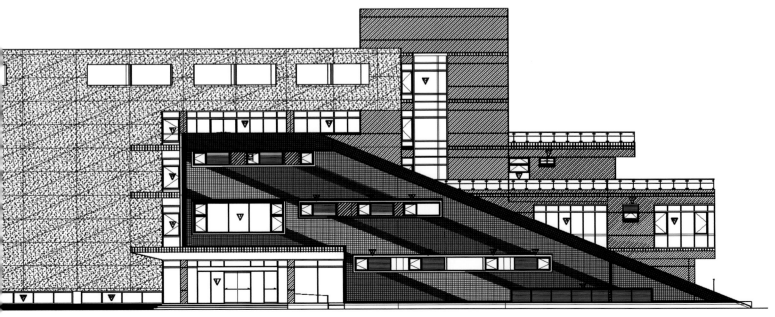

西向立面

Location
Taiwan, China

THSR Branch of Jungli City Kindergarten + Ching Chih Children's Park

中坜市立幼儿园高铁分班 + 青芝儿童公园

Landscape Architect
CTLU Architect & Associates

Client
Jungli City Office

Area
980 m²

Children's Eco-Education Castle
Excellent orientation & opening, greening design, balanced earthwork, eco-education roof... such techniques make it become the most compact, lowest cost "Diamond grade" green building in Taiwan, China.

Children's Exploration Castle
Kids love galloping in open spaces; they also like corners with a sense of security, like a secret base. In the limited space, the design creates a variety of spaces to let children explore its infinite possibilities.

Children's Learning Castle
The classroom is equipped with storage spaces, advanced whiteboard, featured reading corner and learning balcony. There's also a variety of spaces for learning and activities outside the classroom.

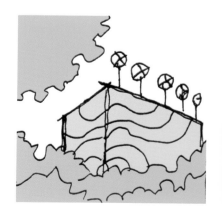

植栽配置图

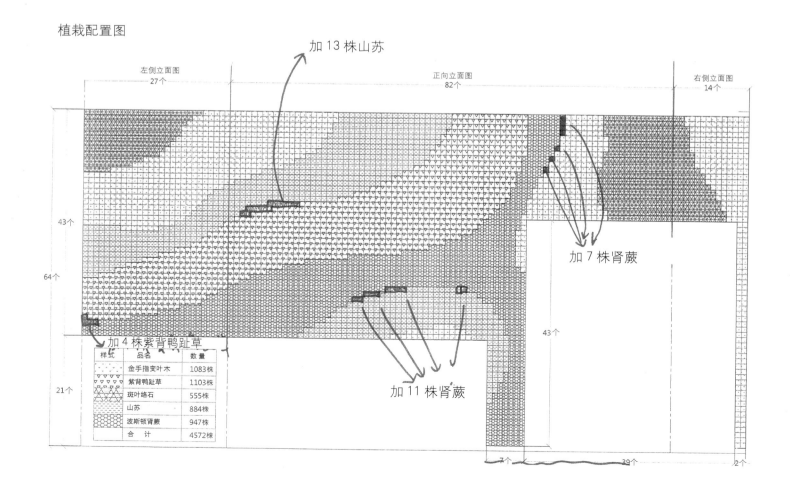

儿童的生态教育城堡

优良的朝向及开窗、大量而丰富的绿化、挖填平衡、生态教育屋顶等手法，使中坜市立幼儿园高铁分班成为台湾最迷你、造价最低的钻石级绿色建筑。

儿童的探索城堡

孩子们喜欢大大的空间可以奔跑，也喜欢秘密基地般富安全感的角落。在有限的面积里，这个设计创造了大大小小的各种空间，供孩子们无限地探索。

儿童的学习城堡

教室单元外多样的角落可供孩子们学习及活动，而教室单元内除了机能性的大量收纳空间、先进的电子白板外，更有别具特色的阅读角及学习阳台。

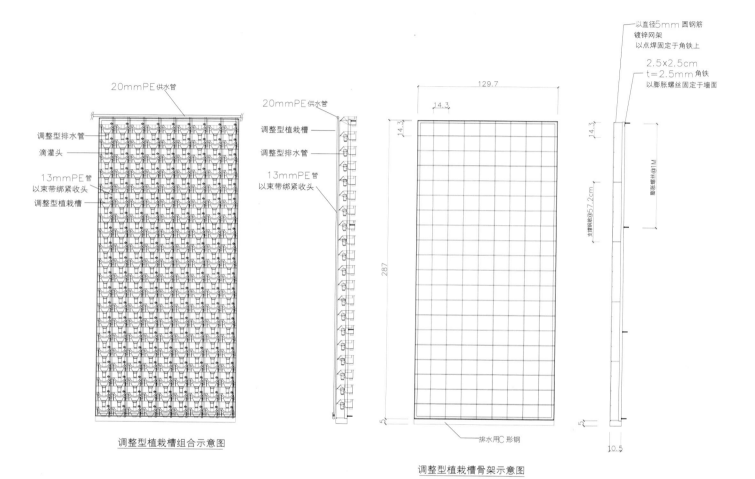

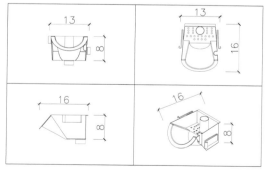

调整型植栽槽单元示意图

植栽网套单元示意图

栽植单元材质：PP 材质

每一栽植单元所需各具独立的灌溉与排水功能

每一栽植单元灌溉后所排出的水分，不得流入其他栽植单元的栽植区，以免土壤病害交互感染

各栽植单元灌溉后之排放水须能汇集至排水沟

每 M2 植栽孔要 44 个以上，植栽 44 株以上

植栽配置图

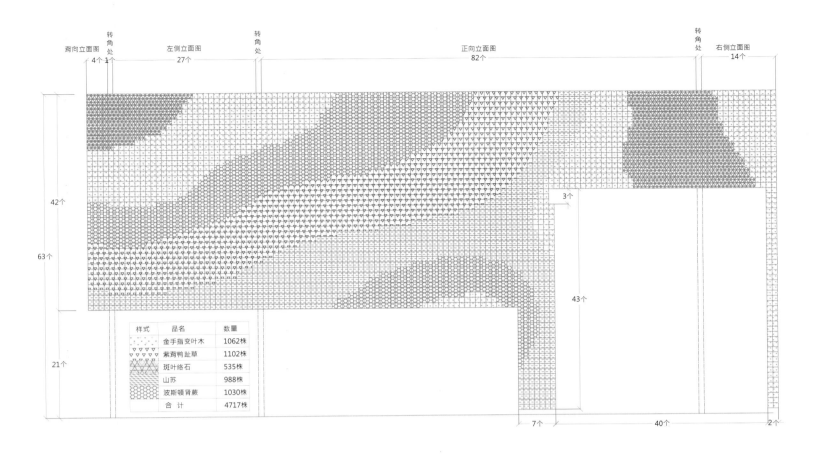

样式	品名	数量
	金手指变叶木	1062株
	紫背鸭趾草	1102株
	斑叶络石	535株
	山苏	988株
	波斯顿肾蕨	1030株
	合 计	4717株

Location
Singapore

Newton Suites
牛顿套房

Landscape Architect
Cicada Pte. Ltd

Building Architect
WOHA

Photographer
Patrick Bingham-Hall

This 36-storey development is a study in environmental solutions to tropical high-rise living. The design integrates several sustainable devices into a contemporary architectural composition, creating a sustainable, contemporary addition to the city skyline.

The building sits at the edge of a high-rise zone and fronts a height-controlled area that affords expansive views of the central nature reserves; a rare luxury in densely built Singapore.

Protruding sky gardens and balconies combined with sunshading screens create outdoor living environments that are sheltered with ample cross-ventilation due to its elevated location and are particularly suited for the hot tropical climate.

Landscape is used as a material – rooftop planting, sky gardens and green walls are incorporated into the design from the very beginning. Creeper screens are applied to otherwise blank walls to create visual delight, absorb sunlight and carbon and create oxygen in the dense environment. Most available horizontal and vertical surfaces are landscaped, creating an area of landscaping that is 130% (110% planted) of the total site.

Trees cover the carpark, project from the sky gardens at every 4 levels and crown the building at the penthouse roof decks. The above ground carpark uses far less energy than an underground carpark and is fully enclosed with creepers, absorbing exhaust emissions. The carpark roof houses a substantial clubhouse with gym, steam room, party areas and a 25m swimming pool with a glass overflow edge.

The end-users experience panoramic views foregrounded by sky gardens and greenery, bringing the indoor-outdoor potential of living in the tropics into the sky, and bringing this to a sector of the community who cannot afford landed housing. Common sky gardens create delight at every lift lobby, turning the wait for the lift in the rush to work into a brief contact with fresh air, trees and sky. The two penthouses include swimming pools with double volume mesh pergolas.

The environmental elements added to liveable apartments and extensive communal areas combine to make a unique tropical building that achieves both Singapore's national vision for a green city and an improved living environment for the inhabitants.

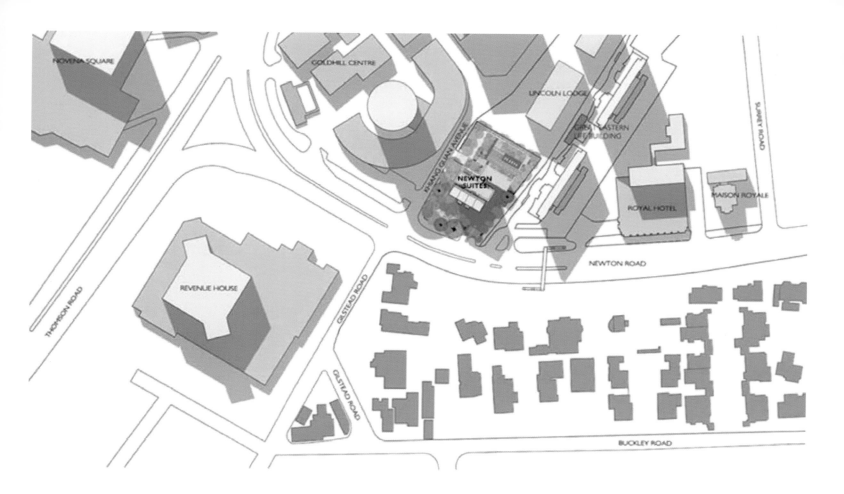

111 NEWTON ROAD
LOCATION PLAN

1:4000
WOHA ARCHITECTS PTE LTD
COPYRIGHT JUNE 2007

　　这栋36层的高层公寓为热带高层生活提供了环境解决方案。项目设计集可持续发展与现代建筑于一体，为现代都市创建一个可持续发展的、现代的建筑样板。

　　本案位于高楼区的边缘，面向限高区，因此视野广阔，可以看到中央自然保护区，在建筑密集的新加坡实属罕见。

　　户外生活环境由突出的空中花园、阳台与遮阳屏幕共同构成，同时因为大楼被升高了位置，所以这里空气流通充分，特别适合于炎热的热带气候。

　　本案设计从一开始就包括了屋顶种植、空中花园和绿墙，我们将景观作为材料。我们在其他的墙上也使用了令人赏心悦目的绿色攀爬植物，这些植物吸收阳光和碳，在人口稠密的环境中释放氧气。我们对大部分可用的水平和垂直的表面都进行了景观设计，整体景观面积是全部面积的130%(110% 为植物)。

　　停车场绿树如荫，每隔四层就有一座空中花园，屋顶公寓平台上也有绿树。地面上的停车场比地下的停车场耗能更少，地面停车场遍布攀爬植物，吸收排放的废气。停车场屋顶有一个内容丰富的俱乐部，包括健身馆、桑拿房、派对区和一个 25 m 长的带有玻璃溢出边缘的游泳池。

　　终端用户体验的是空中花园和绿色植物构成的全景，为热带地区高层住宅开发户内 – 户外的生活潜力，为不能购买接地住宅的人群带来大自然的绿意。空中花园给每个电梯间带来愉悦的风景，当急于上班的人们在等电梯时，他们也可以短暂接触到新鲜的空气，看到绿树和蓝天。两座顶层阁楼都有游泳池，泳池有双倍容量的网格藤架。

　　环境因素、宜居公寓和广泛的公共空间共同构成了一座独一无二的热带建筑，既满足了新加坡建设绿色城市的国家愿景，也为居民提供了一个更优质的生活环境。

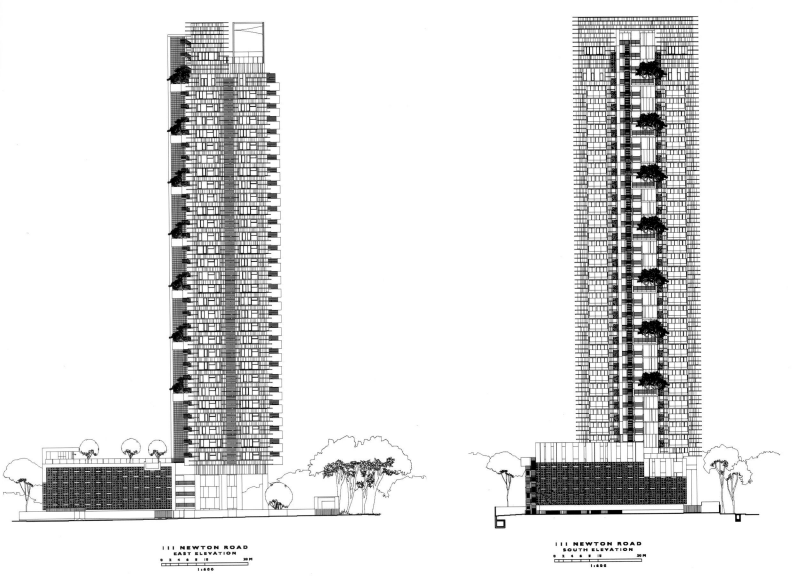

111 NEWTON ROAD
EAST ELEVATION

111 NEWTON ROAD
SOUTH ELEVATION

Location
Johor, Malaysia

Factory on the Earth
地球工厂

Landscape Architect
Junichi Inada

Firm
WIN Landscape Planning & Design Pte Ltd

Building Architect
Ryuichi Ashizawa

Construction
Nakano Construction sdn Bhd

Client
JST Malaysia

Area
46,489 m² (Site Area),
27,143 m² (Landscape Area)

Photographer
Kaori Ichikawa

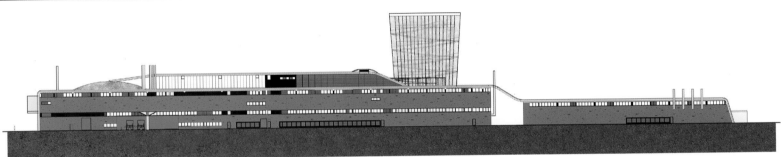

EAST ELEVATION

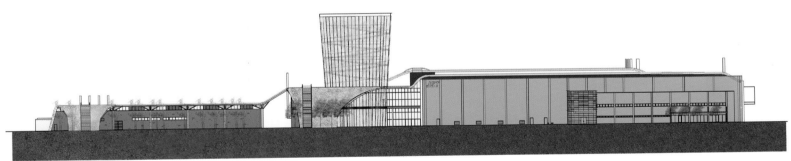

WEST ELEVATION

The factories in the 19th century gave priority to rationality and productivity, so we wanted to transcend the factory typology by incorporating elements that would make the workers proud of the new working environment they would be facing.

Using the power of nature like rain water, sunlight, the wind, geothermal heat and vegetation, we wanted to minimize the production of harmful low carbon expelled to the environment, making the building a sustainable factory.

The plan intends to create a large green roof continuous with the ground extending the earth surface covering the lower functions and spaces.

The roof soil works improve greatly the insolation efficiency of the factory space. The space below is structurally arranged by a forest of hexagonal pillars with a star-shaped top derived from the arabesque patterns, a reference to the surrounding jungle. Rain water that pours down the rooftop slope is pulled into an underground water storage tank through the pipes embedded in the pillars, being used cyclically for plants watering. When flowing into the pond and the wind blows, the rain water that pours down over the rooftop slope, brings a cold breeze to the transitory space between exterior and interior under the roof.

To reduce artificial light as much as possible, the factory is designed to reflect the light that comes from above. Guided by computer simulations, we could predict the amount of skylight reflected and diffused by a reflection panel that shares its shape with the arabesque patterns.

A multistory building houses the offices. This building aligned with the east-west axis allows for a minimized effect of a solar radiation projected on its outer wall surface. The slab along the perimeter of the high-rise building is a slope that connects with the ground level, a continuous walking path for the workers to practice exercise and improve their health.

The façade is provided with a system of wires with vines, which shield the building from solar radiation by sharping a vertical green wall. Natural ventilation is carried through the multistory building to the lower spaces thankfully to its height, creating a current of air.

19世纪的工厂以条理和效率为先，所以我们希望通过各种元素的融合来超越这种工厂模式，将其打造成为一处让工人们引以为豪的全新工作环境。

通过利用自然能源，像雨水、太阳能、风能、地热和植被，我们将最大程度地减少环境中的碳排放，使这个建筑成为一个可持续发展的工厂。

本案旨在设计一处大型绿化屋顶，绵延直至地面，使绿化范围得以扩展，以覆盖低层的功能区和空间。

屋顶种植土的工程结构极大地提高了工厂空间的日晒效率。由六边形柱子构成的柱网支撑起了下层空间，柱子上部被设计成星形，该形状源于阿拉伯式花纹样式，同时也与周边的丛林环境相呼应。从屋顶斜坡落下的雨水通过嵌在柱子中的雨水管流入地下储水罐中，收集来的雨水可用来灌溉植物。当雨水流入池塘时，水流从屋顶斜坡倾泻而下，和着恰好吹来的微风，给位于屋顶之下的室内外之间的过渡空间带来一丝清凉。

为了尽可能减少人工照明，本案设计通过反射将顶部的光线引入到室内。通过计算机模拟，我们能够预测自然光线反射量以及反射板的散射情况，反射板的形状也采用阿拉伯式花纹样式。

办公室位于一栋多层建筑中，大楼沿东西向矗立，使照射在外墙表面的阳光辐射减到最小，高楼四周是一条斜坡的石板路，一直延伸到地面。工人们可以在这条步行道上散步锻炼，强健体魄。

建筑物表面是藤蔓缠绕的金属丝网，构成了一面绿墙，为建筑遮挡日光照射。建筑里因高度关系自然通风气流强劲，能从整个多层建筑中穿过，直到建筑物的下方。

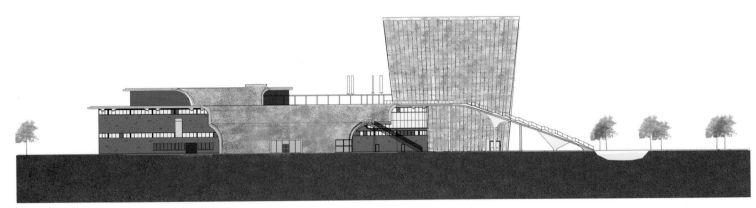

West Elevation

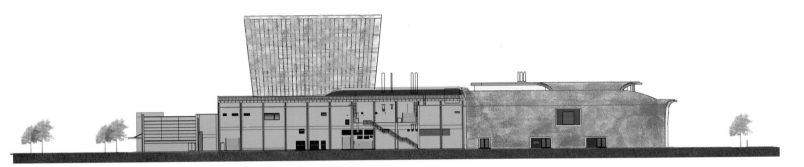

East Elevation

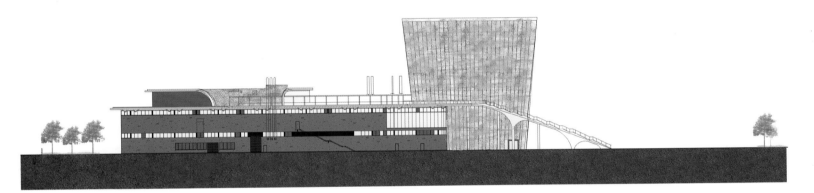

West Elevation

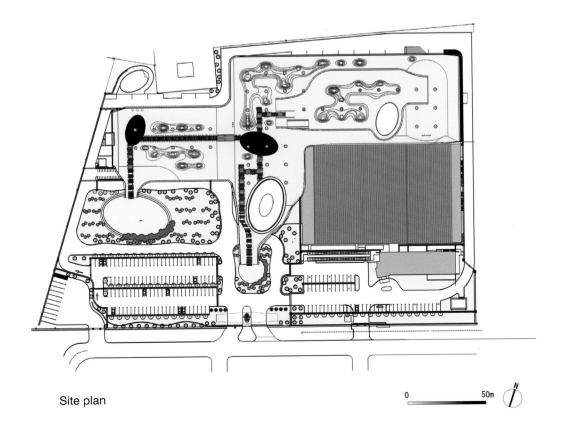

Site plan

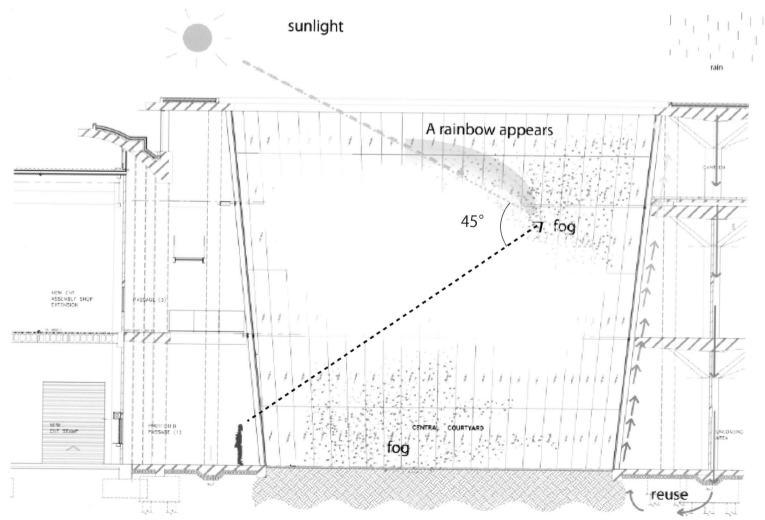

Rainbow Diagram

SUNLIGHT, WIND, WATER

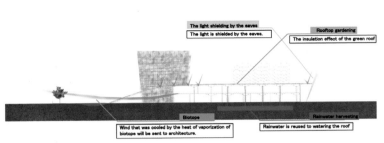

SUNLIGHT, WIND, WATER

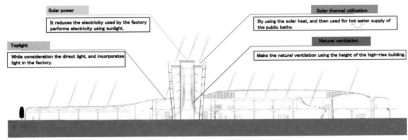

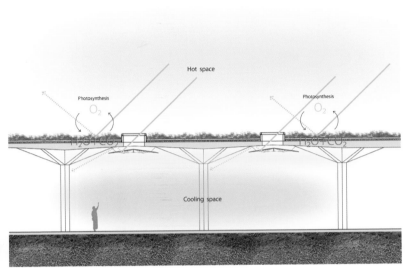

Factory Roof Light Diagram

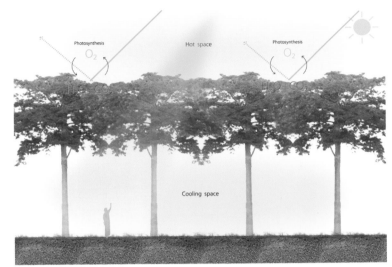

Forest Light Canopy Diagram

Location
Melbourne, Victoria, Australia

Medibank Building
墨尔本医疗福利办公大楼

Landcape Architect	Constructor	Photographer
Hassell	Fytogreen Australia	Fytogreen Australia

Rising 18 levels next to Southern Cross train station, this building was the first of its type in Victoria. Covered with green facades, green roofs and vertical gardens, the Medibank building would set a new benchmark on green building. Fytogreen was contracted to provide research, develop the design and construct the green facades, green roofs and vertical gardens.

Vertical Gardens - 400 m²

Medibank was the primary tenant for the new 720 Bourke Street development in Docklands, Melbourne. Medibank's senior manager asked for a building with significant foliage adorning the facades of the building.

The design company, Hassell, developed the concept of 2 x 200 m² vertical gardens flanking the 720 Bourke Street entrance on the south east of the building. Fytogreen's botanist used 72 species and 11,600 plants to create this design. Fytogreen pre-grew these vertical gardens in 5 months prior to installation in shade houses. This southeast-facing wall only got a small amount of early morning sun and the rest of the day it's in shade.

Green Facades - 1326 m²

The green facades were made up of 520 planter boxes each pre-built and pre-grown at Fytogreen's facility in Somerville, then transported and craned into position from level 1 up to level 16. Each planter box had the dimensions of 600 mm (high) x 600 mm (wide) x 600 mm (deep). The bottom 1/3 of the climbing frame was attached to the planter box during pre-grown and then was attached to the rest of the climbing frame when the planter box was positioned on the building. 6 species of climbing vines were selected from our research plots, which included Trachelospernum jasminoides, Pandorea pandorana, Pandorea jasminoides, Muehlenbeckia adpressa, Muelenbeckia complexa and Hibbertia scandens. These were planted and pre-grown for 3-6 months prior to installation. The green facades adorned 3 aspects of the Medibank Building and would need to survive challenging and wide range of conditions from level 1 up to level 16.

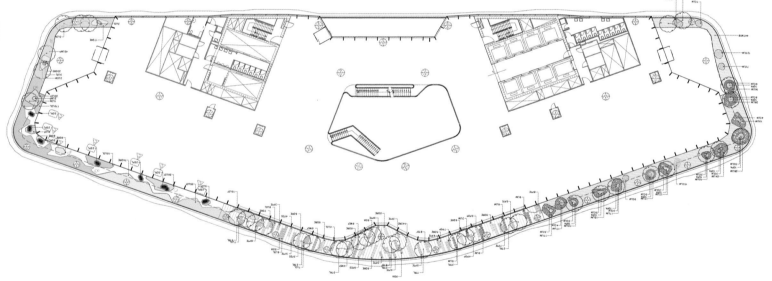

项目大楼位于南十字星火车站旁，18层高，是维多利亚州第一座此种类型的建筑。建筑立面和屋顶都是绿色的，配备有垂直花园，墨尔本医疗福利办公大楼将成为绿色建筑的新标准。Fytogreen公司签约负责设计的调研、开发及绿色外墙、屋顶和垂直花园的建设。

400 m² 的垂直花园

本案中的医疗福利办公大楼是墨尔本达克兰区新Bourke大街发展区720号的主要租户。医疗福利办公大楼的高级经理要求用树叶饰物来装饰建筑外表。

建筑公司Hassell的构想是，在Bourke大街720号建筑物东南的入口两侧设计两个面积为200 m²的垂直花园。Fytogreen的植物学家使用了72种不同的种类、11 600株植物来完成垂直花园的设计。安装前5个月，Fytogreen提前在遮阳棚里种植垂直花园。东南向的墙只能在早上得到一点日照，其他时间都处在阴影中。

1 326 m² 的绿色外墙

本案绿色外墙由520个花槽构成。这些花槽是在Fytogreen公司位于Somerville的工厂提前制作并种植的，被搬运过来后，被吊车分别送至一层到十六层。每个花槽的体积为600 mm×600 mm×600 mm。在种植前期，花槽附着在攀爬架的下三分之一处，当花槽在大楼中放置好后，植物就爬满了攀爬架的其余位置。我们从研究中心选择了六种攀爬藤蔓，包括络石、粉霄藤、粉花凌霄、孔雀草、丝蔓和束蕊花。这些植物在安装前3~6个月被提前种植。医疗福利办公大楼三面均有绿色外墙装饰，这些绿墙植物需要适应从1层到16层的不同环境。

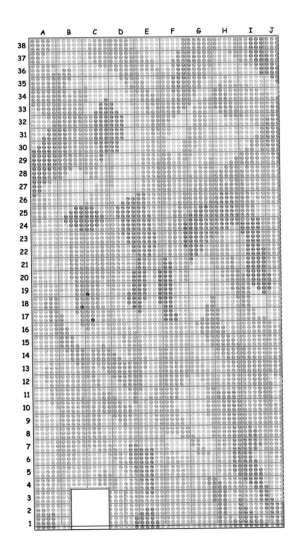

Location
Amiens, France

The Picardy Regional Chamber of Commerce and Industry
皮卡地区商会

Landscape Architect
Chartier-Corbasson Architectes

Client
CRCI de Picardie

Photographer
R. Meffre & Y. Marchand

The Bouctot-Vagniez Town Hall in Amiens is a remarkable building, an architectural testament to the glories of nineteen-twenties Art Nouveau. Our project is concerned with designing an extension to this unique building, which is home to the Picardy Regional Chamber of Commerce and Industry.

All the essential features of the project are represented in a plinth of living greenery that creates a link between the new wing, the existing premises and the gardens. The offices will be situated above this greenery plinth. They are housed in two separate spaces divided from one another by an atrium that will allow natural light and air to penetrate the heart of the building. Screen-printing technology protects certain perspectives by shading the glazed areas or leaving them clear, according to the needs created by the utilization of the rooms behind. To the south, on the roadside elevation, a double skin of metal mesh allows for ventilation and creates a sunscreen, creating a secluded atmosphere in the offices.

As regards the garden elevation, the design forms part of the existing landscaping as a sort of kink in the boundary wall. The hall opens out as broadly as possible onto the gardens, and the ground floor rises up to embrace a wide panoramic bay window creating a fluid, light-filled space.

The offices are divided into two volumes:

- The annex volume focuses all service areas: toilets, copiers, cloakrooms, corridors. The staircase block services are positioned to the front of it, exploiting the immediate view of the hotel.

- The main volume welcomes office floors, between the avenue and the garden.

A rift full height between these two volumes allows light and air in the heart of the building. Oriented on south tower of the hotel Bouctot-Vagniez, this rift is crossed by glass bridges leading to offices in the service areas. The view from the avenue is possible to the hotel, linking the two buildings, the existing and extension into one project.

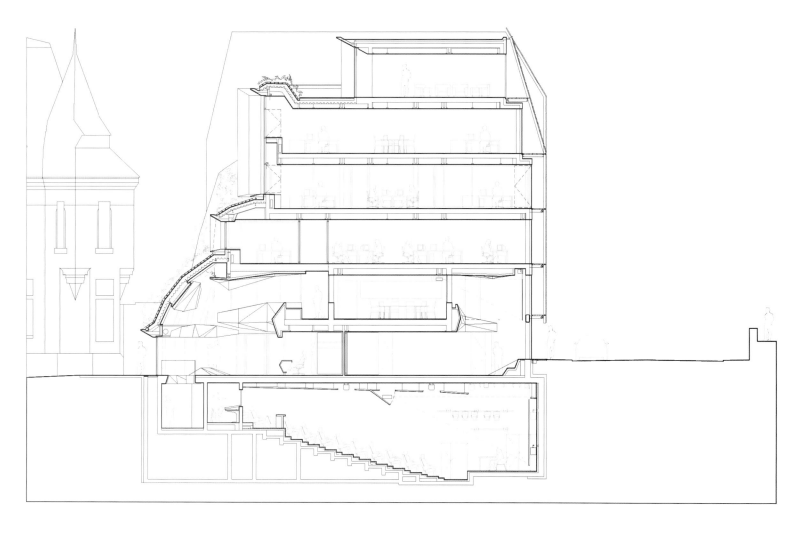

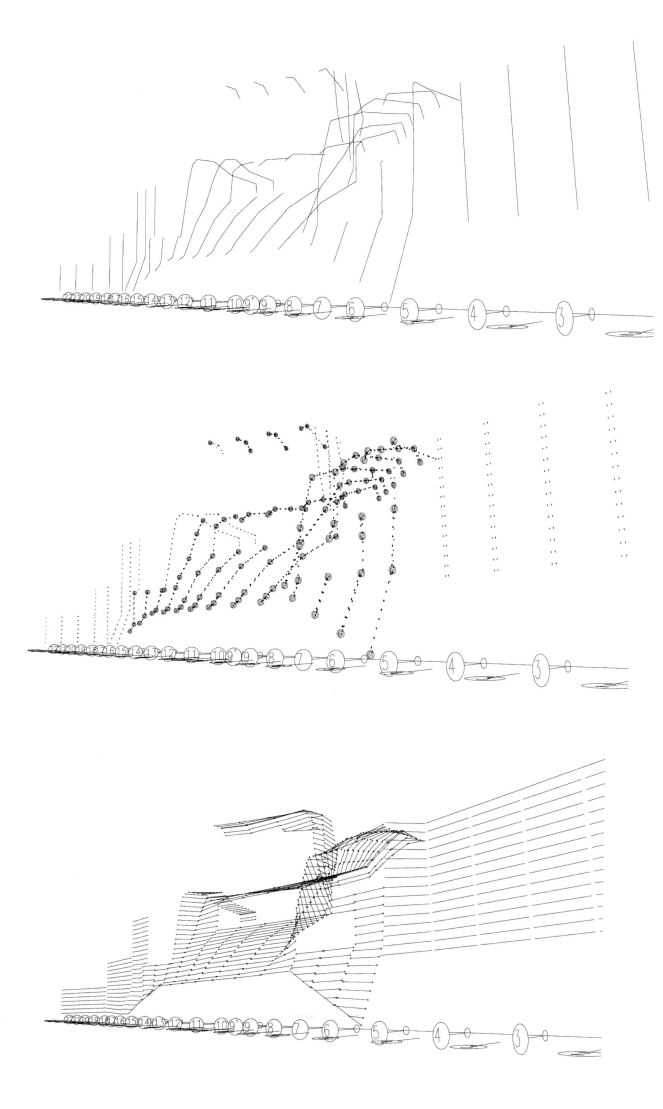

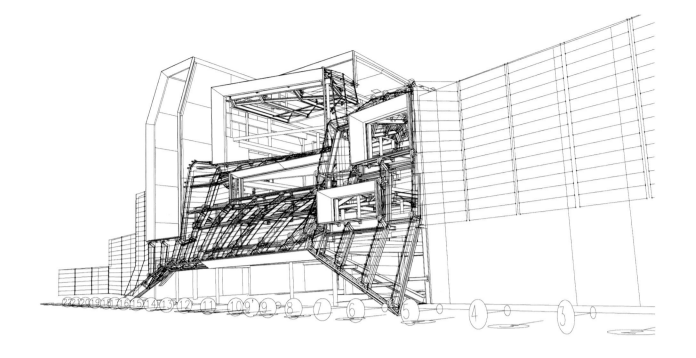

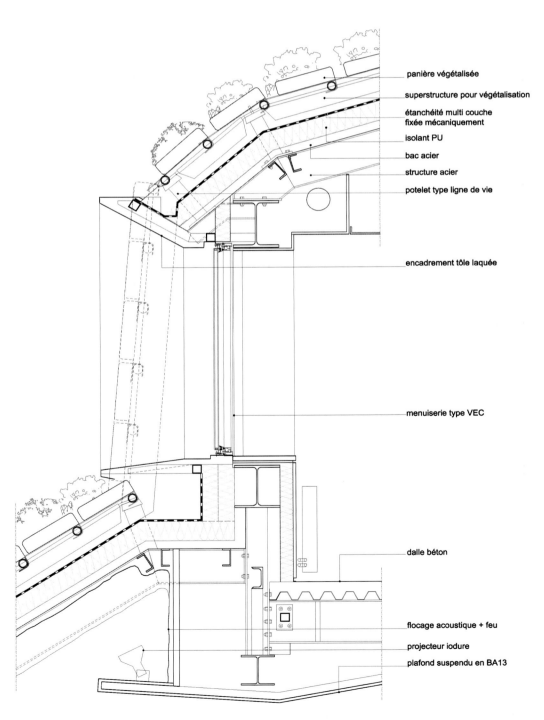

亚眠市的 Bouctot-Vagniez 市政厅大楼别具一格，该建筑见证了 19、20 世纪新艺术的辉煌。本案是为这座独一无二的建筑设计的一个扩展部分，这里也是皮卡地区商会的所在地。

本案核心特色展现于生机盎然、布满绿色植物的基座，基座连接新的楼翼、现存的地基和花园。办公室都位于绿色基座上，分布于由一个中庭隔开的两处空间，自然光和空气可以通过中庭直抵建筑中心。

本案根据后面的房间利用率的需要，采用丝网印刷技术通过遮挡釉面区或留出空当，来保护某些视角。南边朝向道路的立面上，双层表皮金属网既可通风，又可遮阳，使办公室氛围安静幽僻。

考虑到花园立面景观，在设计中将现有景观在边界墙上做了一定的扭曲。大厅尽可能地面向花园敞开，抬升一层以容纳一个宽阔的全景飘窗，享受一个流光溢彩的空间。

办公室分为两部分：

- 配楼主要是服务区：厕所、复印室、衣帽间、走廊。楼梯区域服务设施设置在前面，这样可以直接看到酒店。

- 主楼为办公区，位于大街和花园之间。

两栋楼之间有一道全高裂隙，使得光和空气可以穿透建筑。裂隙面向 Bouctot-Vagniez 酒店南部塔楼，玻璃廊桥穿过裂隙通往服务区的所有办公室。从酒店可以看到林荫大道，空间连接了两栋楼，将现存的建筑和扩展建筑融为一体。

Location
Amsterdam, The Netherlands

Sportplaza Mercator, Amsterdam, The Netherlands
荷兰阿姆斯特丹墨卡托体育馆

De Baarsjes in Amsterdam is a multicultural neighborhood that is home to people from 129 different countries. The city district wants to boost community life in the neighborhood. The authorities therefore choose a building which combines swimming pools, a therapy pool, fitness, aerobics, a sauna and steam bath, a party centre, café and childcare alongside a fast food restaurant (jobs for the unemployed in the neighborhood). Each individual element attracts different target groups, so the entire population will be able to use it in the end. Inside, everyone can see other activities, intriguing their interest and inspiring them to use other facilities as well. Because the building is constructed in a park, people living nearby it request that it should be as green as possible; we completely cover it in vegetation.

Now, with its green façades and roof, Sportplaza Mercator marks the start and end of the Rembrandtpark. From a distance, it seems like an overgrown fortress flanking and protecting the entryway to the 19th-century city. Glimpsed through the glass façade, a modern spa-style complex glistens, completed with swimming pools, fitness space, and restaurant and party facilities. The entrance seems like a departure hall from which the various visitors can reach their destination.

The building is designed as a city – a society in miniature – inside a cave. The building is full of lines of sight and keyholes that offer perspectives on the various visitors, activities and cultures in the building. Sunlight penetrates deep into the building's interior through all sorts of openings in the roof. Low windows frame the view of the street and the sun terrace.

Building Architect
VenhoevenCS architecture+urbanism

Area
7,100 m²

Photographer
VenhoevenCS architecture+urbanism, Copijn en Luuk Kramer

阿姆斯特丹 De Baarsjes 社区是一个多元文化社区，居住着来自 129 个不同国家的居民。市区想要改善附近的社区生活，因此，有关当局选择建造一座建筑物，内设游泳池、疗养池、健身房、桑拿和蒸汽浴、派对中心、咖啡馆和儿童托管中心、快餐店（为社区失业者提供工作）。每个元素吸引不同的目标群体，因此最后全部人员都能使用这个建筑。在里面，每个人都可以看到其他的活动项目，由此引起他们的兴趣，鼓励他们也使用其他的设施。由于这个建筑位于公园中，住在附近的人们要求它尽量呈现绿色，因此建筑师让建筑披上了厚厚的绿色植被外衣。

现在，凭借其绿色的立面和屋顶，墨卡托体育馆可以算作是伦勃朗公园史无前例的巅峰之作。从远处看来，它好像一个蔓生的城堡，作为侧翼，保护着这个 19 世纪城市的入口通道。从玻璃立面望去，这个现代的带有游泳池、健身空间、餐厅和聚会设施的温泉型综合建筑熠熠生辉。其入口处看上去就像一个机场候机楼，形形色色的游客们由此抵达他们的目的地。

这个建筑物被设计成了一个洞穴之中的城市——一个社会的缩影。建筑物内不同游客可以从不同视角看到大楼里开展的各种各样的活动和文化景观。屋顶天窗让阳光深入建筑内部。低处的窗户框住街道和阳光露台的景致。

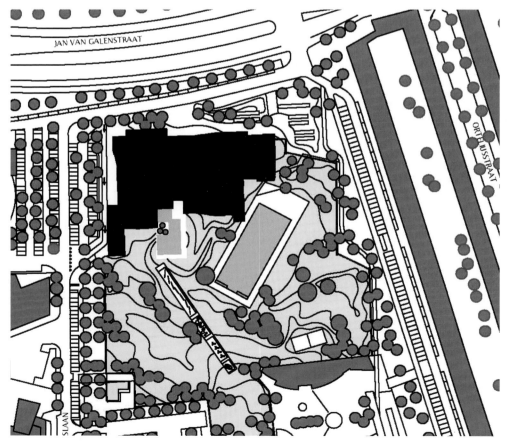

SITE PLAN

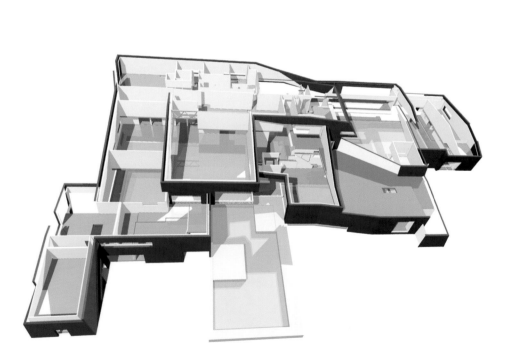

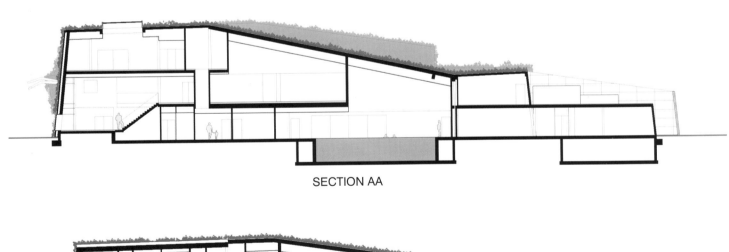

SECTION AA

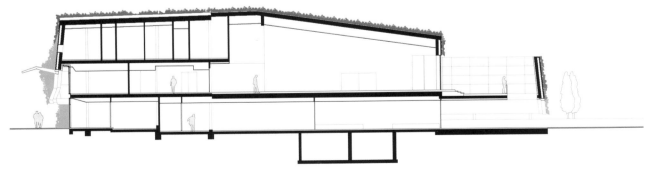

SECTION BB

Sportplaza Mercator

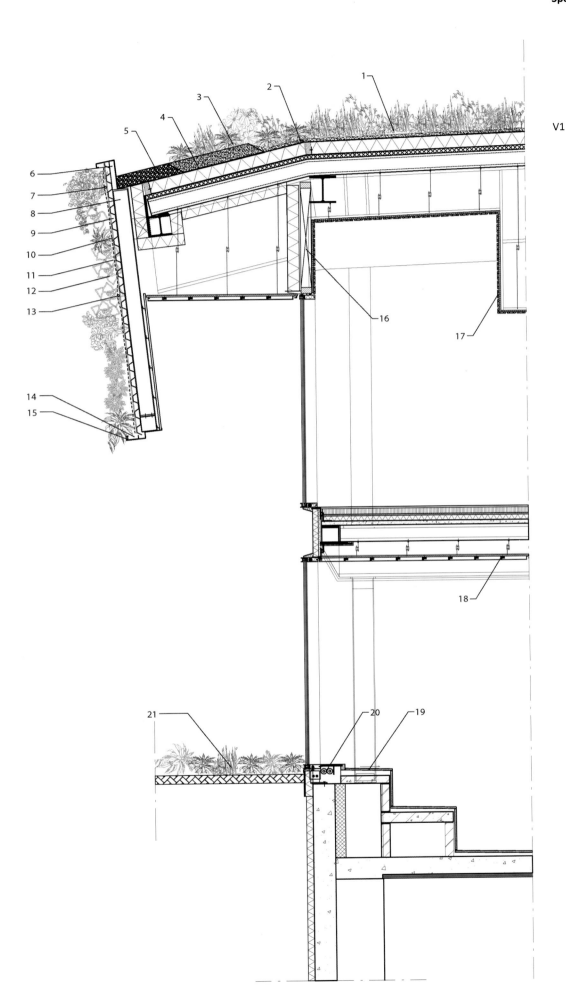

Legend Details

Scale 1:20

V1 Sectional detail V1

section of green elevation

1. sowed sedum grasses
2. 40 mm substrates
3. permanent plants 6 per m^2
4. 150 mm substrate
5. 150 mm gravel
6. tube 60x60x6 mm
7. water system, drip pipe
8. columns green façade
9. sheet-pile wall
10. synthetic plate 10mm
11. 3 layers of textile & foil 10mm
12. Plants
13. water system, drip pipe
14. waste pipe
15. steel frame
16. wooden frame construction
17. wooden lath
18. expanded metal ceiling
19. Ivy plants
20. Convector
21. Granite floor

Location
Singapore

NEX Shopping Center

NEX 购物中心

Landscape Architect
Broadway Malyan

Building Architect
SAA Architects

Text
Robert Such

In most shopping centers, shoppers are encouraged to move around. At NEX, that is also true, but it offers something different as well—a number of green plants break out spaces in which to sit and take a breath. NEX also has to be designed to integrate some 56,000 square meters of retail floor space, a cineplex and a public transport hub.

Located in Serangoon in north-east Singapore, the seven-storey NEX shopping center is bordered by three main roads and surrounded by medium and high-rise Housing Development Board (HDB) residential areas.

NEX's principal design features include a landscaped roof terrace and a series of linked landscaped decks and void spaces. Found at different levels throughout the building, these green spaces provide a datum to various parts of the mall internally, and externally break down the urban scale of the building. They also serve as way-finding points and as places for people to rest and to socialise.

The rooftop, with its various public amenities and social spaces, such as children's playground, public library, dog walking enclosure and education centres, is designed like a town square around a courtyard garden. NEX's open-air facilities also serve to supplement those found in the local neighbourhood.

A series of escalators and walkways connects the roof to the landscaped decks, car park, pedestrian link bridges and transport hub (bus interchange and underground MRT station). Forming an external pathway that starts from where a civic plaza will be built, they permit access to the interior at each level, and as people use the escalators, the movement animates the outside of the building.

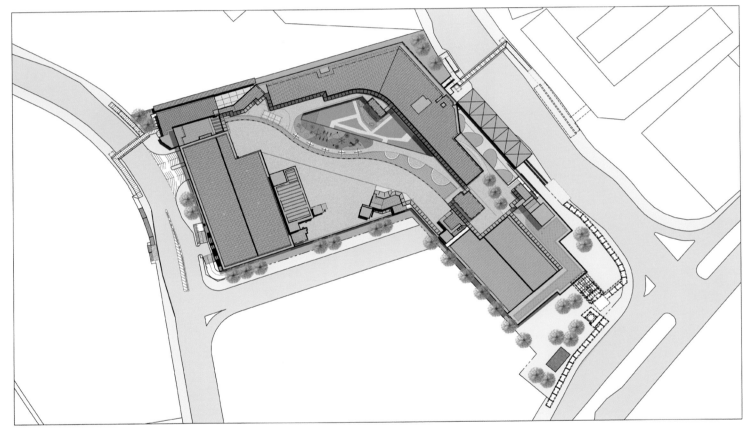

SITE PLAN

在大多数购物中心，商家都会鼓励消费者四处逛逛。在 NEX 购物中心也是如此。但不同之处在于，NEX 还为消费者提供了一系列的绿色休息区，可供人们歇息。NEX 购物中心还包括了 56 000 m² 的零售区、一家电影院和一个公交枢纽。

NEX 购物中心共七层，位于新加坡东北部的实龙岗，三面临街，周围中高层的住宅区林立。

NEX 购物中心的主要设计特色包括景观屋顶平台，一系列相连的景观甲板和上空空间。遍布于各个楼层的绿色空间成为联系商场内部不同部分的交叉点，并从外部将建筑空间进行了分割。这些绿色空间同时还是人们的路标，供人们休息和社交。

本案屋顶有各种公共设施和社会空间，如儿童游乐园、公共图书馆、遛狗围场和教育中心。屋顶设计得像一座围绕庭院花园而建的城市广场。NEX 的户外设施也可以补充当地小区设施的不足。

一系列扶梯和通道连接屋顶到景观甲板、停车场、人行天桥和交通枢纽（公交和地铁换乘站）。这些扶梯和通道构成了一条外部通道，可以从一座将建的公共广场进入内部的任何一层，当人们乘坐扶梯时，扶梯的上下运动使整栋大楼显得富有活力和生机。

GREEN WALL ELEVATION

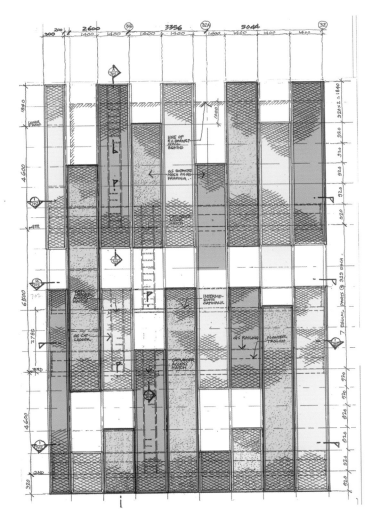

GREEN WALL SECTION

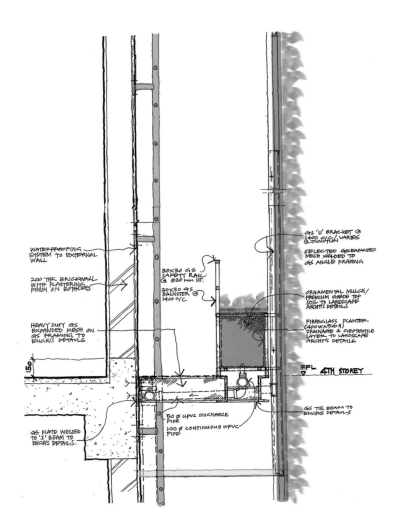

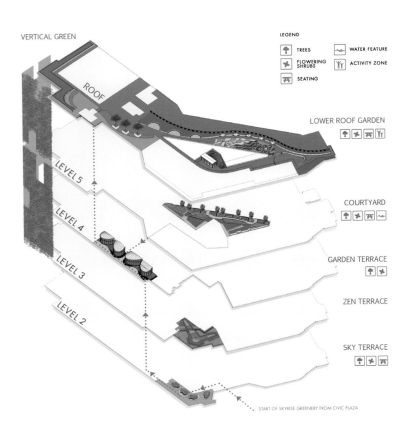

Location
Singapore

Ocean Financial Center
海洋金融中心

Landscape Architect
Tierra Design (S) Pte Ltd

Building Architect
Architects 61 Pte Ltd

Area
6,109 m²

Photographer
Amir Sultan

The tropical gardens in the sky at the Ocean Financial Center are designed within the interstitial spaces between the building structure and façade framing. Creating a connection to the natural environment, the vertical landscaping at the highest floors from the 39th to the 43rd level provides a lush setting, providing much needed relief to users who spend most of their time in the artificially ventilated offices. The canopy trees at the 41st floor create an interesting play of scale through different textures and colours of planting. The vertical green frames are comprised of planters located at every 1.5m, planted with fast growing Thunbergia grandiflora climber species to ensure a full green cover. A combination of profuse flowering plants and colourful textured leaf foliage of shrubs at the base of the green columns lend an exciting and vibrant feel. Auto-irrigation & fertilisation system installed for all the planter areas ensures that through timer-controlled drip nozzles, slow, steady and precise amounts of water and nutrients are provided. Rain sensors are installed at planters to cut off irrigation during rainy days. All these measures ensure that water is judiciously used in just the required amounts for the plants to thrive and sustain the lush urban high-rise gardens at the Ocean Financial Center.

At the ground level, a large car park has been built to accommodate visitors and tenants at OFC. Its unfortunate location means that it jars against the luxurious entrance that has originally been envisioned. The designers decides to change the way the car park is perceived by turning it into a tourist attraction itself. Now, the car park hosts the world's largest green wall. Looking up at it, it is hard to say what lies behind. The walls showcase beautiful maps of the region and the world, all in varying shades of green, and at night, the lighting highlights the beauty of the walls.

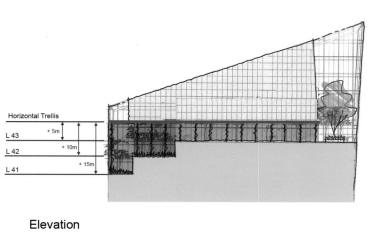

Elevation

Trellis roof plan (L41-42-43)

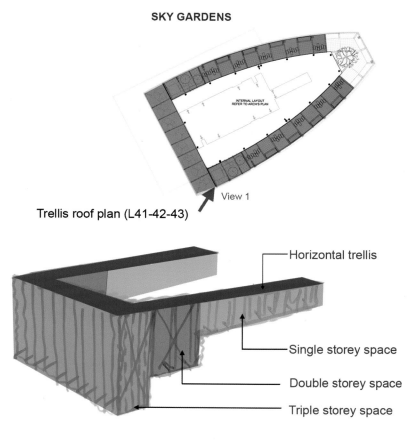

View 1

Schematic diagram

海洋金融中心的空中热带花园被设计在建筑结构和立面框架的间隙之中。为了与自然环境相联系，从39层到43层的最高层的垂直景观为建筑物提供了苍翠茂盛的背景，给大部分时间都待在靠人工通风的办公室里的用户提供了很多必需的安慰。在第41层的树冠通过不同植物的不同纹理和颜色打造了一个有趣的玩耍之地。垂直的绿色框架由间隔1.5 m的花架组成，所种的是一些生长快速的山牵牛攀缘类植物，以确保全面的绿色覆盖。大量的开花植物加上在绿色柱子底部的彩色纹理叶片灌木植物给人以兴奋和生机勃勃的感觉。在所有的种植区域安装自动灌溉和施肥系统，通过时间控制水滴的喷嘴，来确保缓慢地、稳定地、精确地给植物提供水分和养分。花架上安装了雨水感应器，在下雨天能够切断灌溉系统。所有的这些措施都是为了保证水被明智而合理地使用，保证植物茁壮成长所需的水量，维持海洋金融中心城市高层花园的苍翠繁茂。

为了满足海洋金融中心的访客和租户的停车需求，一个大型的停车场已经被建在一层。令人遗憾的是，它的位置与最初设想的OFC的豪华入口相冲突。设计者们决定一改停车场在人们心目中的形象，将停车场本身设计为一处观光胜地。现在，该停车场拥有世界上最大的绿墙。抬头望去，人们很难说出其后面是什么。该墙展现了区域和世界范围内的美景地图，在绿色的阴影里变化万千。在夜间，照明设备更加凸显了墙体的美丽。

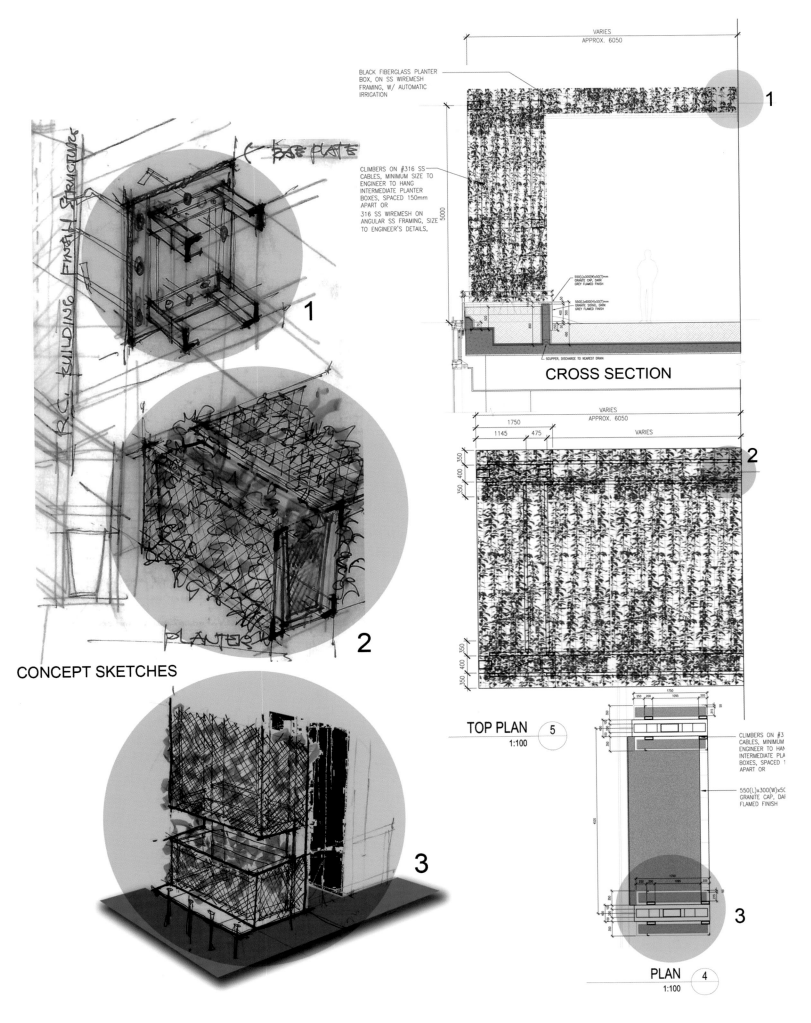

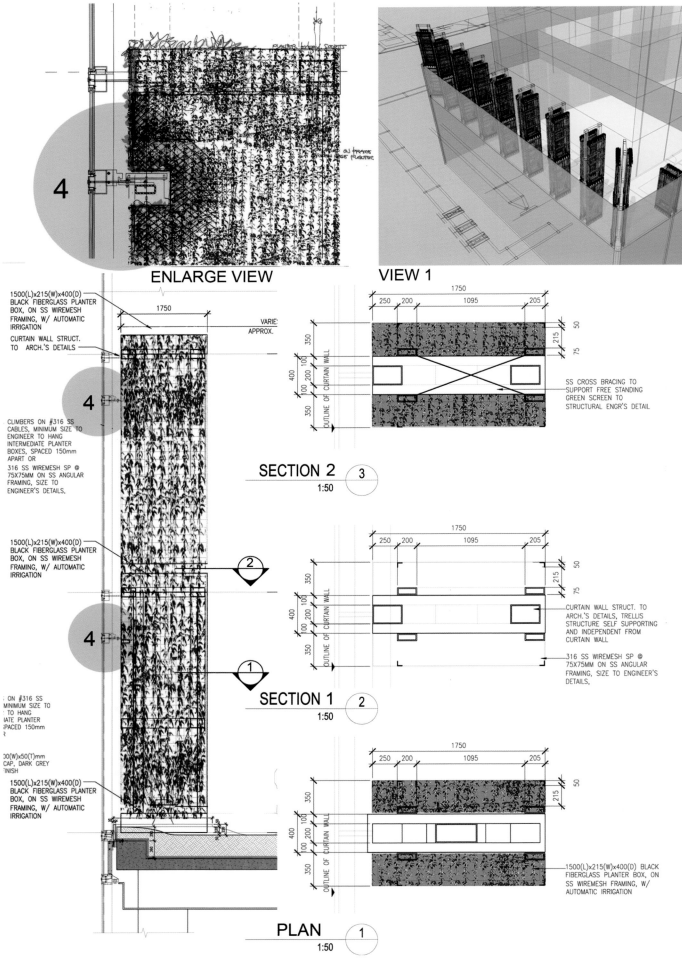

Location
Makati, Philippines

Gramercy Sky Park
葛莱美西空中公园

Landscape Architect
Pomeroy Studio Pte Ltd

Building Architect
Jerde Partnership Inc., Pomeroy Studio Pte Ltd

Photographer
Tom Epperson for Century Properties

Pomeroy Studio wished to evoke the sense of the original Gramercy Park in Manhattan as a recreational social space that could also offer micro-climate benefits to its inhabitants and visitors alike. The original pocket park in Manhattan had a lush character in the context of the city, with mature trees and pleasant pathways in which residents can relax, exercise or engage with others.

The design intent therefore sought to recreate the social and spatial experience of the original park setting, albeit replicating its essence vertically. Mapping the spaces of the original Gramercy Park helped the design team to identify pathways, seating areas, trees and the fountain, and allowed the characteristic social and spatial elements to be superimposed in plan as reference points. The greenery was then objectively studied using the green plot ratio method to identify the areas of dense foliage and sparse foliage, thus allowing for not only a visual reference point of its greenery, its tones and textures, but also its density based on the leaf area index. This further allowed the horizontal greenery of the original Gramercy Park to be extrapolated and rotated in 90 degrees to effectively create a green wall "mural" within the Gramercy Sky Park. The curvilinear green mural alluded to elements within the original Gramercy Park whilst avoiding pastiche.

Potted indigenous species, suspended and drip-irrigated, were used to clad the walls and to create the green mural to evoke the sense of a "hanging garden", with the beams cladded in mirrors to further reflect and therefore to accentuate the greenery. The hanging garden theme was continued through the central void space to create the tallest green wall in the Philippines. The net effect of the vertical greenery within the void space at the center of the tall building was not only visual but also embodying important spatial, social and environmental properties in a city characterized by high pollution levels, humidity and heat.

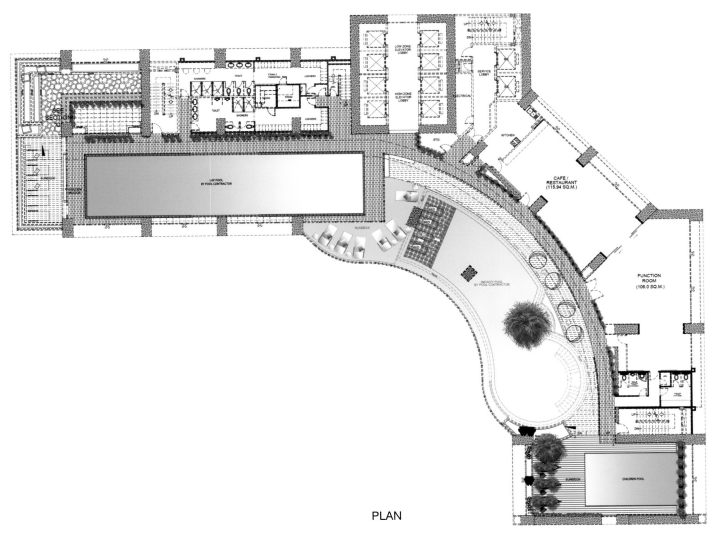

PLAN

Green Wall Irrigation and Fertigation System

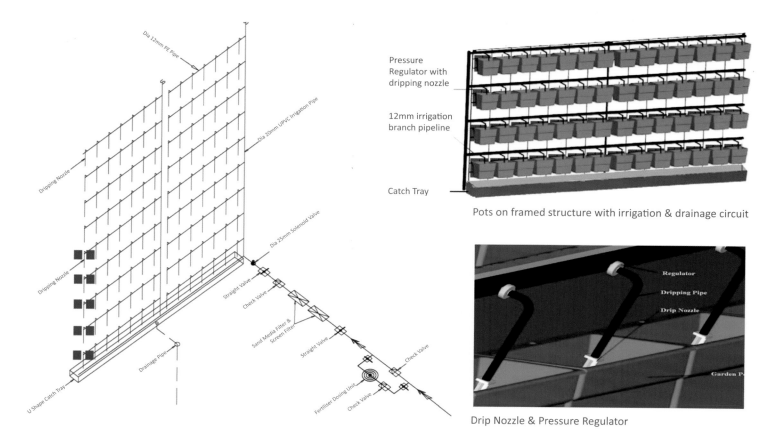

Pots on framed structure with irrigation & drainage circuit

Drip Nozzle & Pressure Regulator

Doubly loaded pots on aluminum frame

Greenwall Elevations

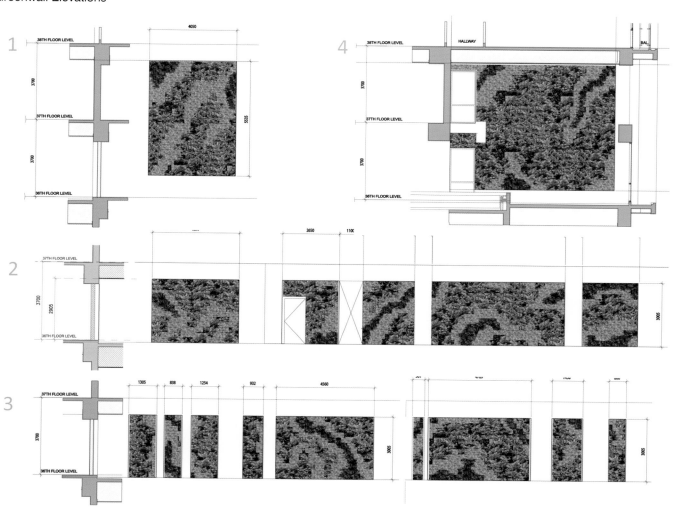

　　Pomeroy 工作室希望人们意识到，曼哈顿葛莱美西原始公园可以作为一个社会娱乐空间，又可以给居民和游客带来小气候的福利。曼哈顿早期的小型花园，在城市脉络中呈现苍翠繁茂的特点，树木葱葱，小径纵横，在这里，居民可以放松身心，锻炼身体，结交朋友。

　　因此设计试图寻求从社会和空间角度重塑人们对原始公园环境的体验，尽管是从垂直角度来实现的。绘制原始葛莱美西公园的空间图可以帮助设计团队确定路径、座位区、树木区和喷泉区，在平面图中使社会和空间元素相叠加，共同作为参考点。对绿色植物进行客观的研究，使用绿色容积率的方法来确定植物的密度和稀疏度，因此不仅要考虑其绿叶、色调和纹理的视觉参考点，还有它基于叶面积指数的密度。这进一步使原始葛莱美西公园水平绿化被延伸并旋转 90°，从而在葛莱美西空中公园中有效地打造出一面绿墙壁画。曲线形的绿色壁画提及到原始葛莱美西公园中的一些元素，但避免了模仿。

　　滴灌盆栽的本地品种，悬浮着覆盖墙壁，打造绿色壁画，来唤起人们对"悬挂花园"的意识。光线照在包覆在柱子外面的镜面上，经过进一步反射，凸显了绿色植物。"悬挂花园"这一主题延续至中央上空空间，打造了菲律宾最高的绿墙。在这座以高污染、高湿度、高热量为特点的城市中，这栋高层建筑中央上空空间的垂直绿化净效应，不仅体现在视觉上，同时还体现在空间、社会和环境属性上。

Location
Birmingham, UK

LG Arena
LG 广场

Landscape Architect
DLA Landscape & Urban Design

Client
The National Exhibition Centre

Living Wall System Supplier
ANS Group

Contractor
GF Tomlinson

Sub-Contractor (Steel Sub-structure)
Lowe Engineering Ltd

As a fundamental part of the successful transformation of the LG Arena, DLA Landscape develops a landscape strategy, which is inspired by the former NEC Arena's long history of hosting live music events. Every aspect of the landscape scheme, from paving and planting design to the external lighting strategy and the living wall reflects a physical expression of sound waves radiating from the Arena out across the public realm. Both the vertical bands of planting and the panels of perforated stainless steel are intended to reinforce this sound wave concept within the wall design.

The wall also has a practical purpose which is to direct visitors towards the arena itself but, as the arrival is an important part of the visitor experience, rather than simply throw up a standard wall we decides to design something beautiful, unique and living – to create a real wow factor for the new arena entrance. The wall is slowly built up in height as you near the entrance to add to the sense of anticipation and excitement of the show to come.

Over 15,000 shrubs, ferns, ornamental grasses and hardy perennials are planted at a density over 100 plants per square metre on the wall, which rises in steps from 3 metres to 5.5 metres. In addition the wall is back-lit in the LG Arena colours to create a striking night-time feature.

The plants are watered and fed via a computer-controlled hydroponic irrigation system housed within the wall structure. The plants have been chosen for their evergreen, shade-tolerant and low maintenance properties, although they will be pruned twice annually to make sure the wall remains in good condition throughout the year.

The irrigation system at LG is fed by mains water supply and whilst it is more sustainable if fed from a rainwater harvesting system which brings with it additional need for filtering the water and increased maintenance.

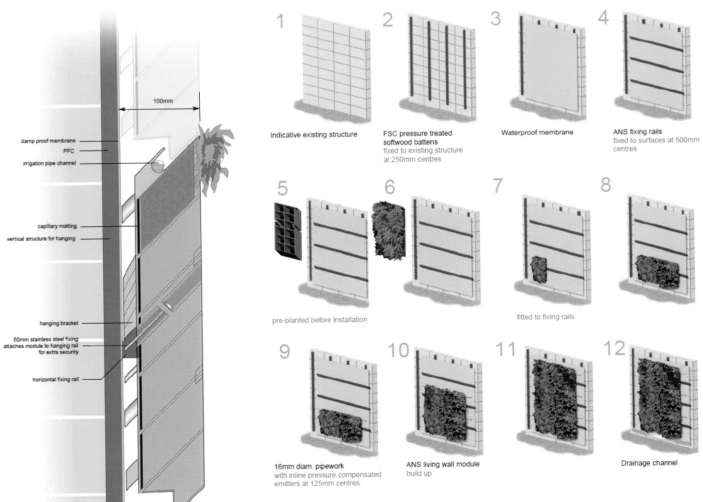

- damp proof membrane
- PFC
- irrigation pipe channel
- capillary matting
- vertical structure for hanging
- hanging bracket
- 60mm stainless steel fixing attaches module to hanging rail for extra security
- horizontal fixing rail

1 Indicative existing structure

2 FSC pressure treated softwood battens fixed to existing structure at 250mm centres

3 Waterproof membrane

4 ANS fixing rails fixed to surfaces at 500mm centres

5 pre-planted before installation

6

7 fitted to fixing rails

8

9 16mm diam. pipework with inline pressure compensated emitters at 125mm centres

10 ANS living wall module build up

11

12 Drainage channel

LIVING WALL

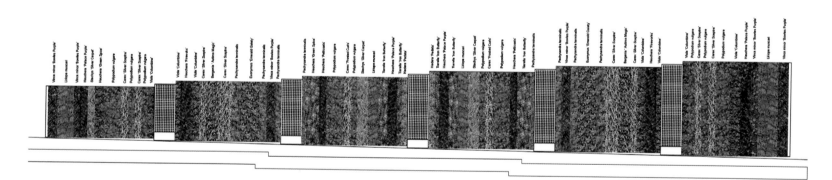

GREEN WALL FINAL

为成功改造 LG 广场，DLA 景观公司开发出一套景观战略。该景观战略受之前的 NEC 广场的启发，NEC 广场举办现场音乐会的历史可谓悠久。从铺设到种植计划，到外部照明策略和活体墙，景观计划的每个方面都反映了声波从广场穿过公共空间传播出去。在墙的设计中，垂直的植物带和打孔的不锈钢板被用来强化声波。

墙有实际功能，可以引导游客来到广场，但是到达目的地是游客体验的一个重要部分，与其简单竖立起一面标准的墙，我们觉得不如设计一面漂亮的、独一无二的垂直绿墙——让游客到达新的广场入口时发出惊叹。当游客走近入口时，墙的高度递增，增加游客对接下来的视觉感受的参与感和兴奋之情。

墙上的植物密度超过每平方米 100 株，包括 15 000 多株灌木、蕨类植物、观赏草和顽强的多年生植物，从 3 m 到 5.5 m 高度不等。此外，LG 广场的缤纷色彩从背面照着垂直绿墙，营造了一种引人注目的夜间特色。

在墙的结构里有一个电脑控制的水耕灌溉系统，可以给植物提供水分和养分。我们依据植物特性来选择这些植物，如常青、耐阴、低维护，不过这些植物需要每年修剪两次，确保墙体常年维持在良好状态。

LG 的灌溉系统来自于自来水供应，如果灌溉水来自雨水收集系统，虽然更加环保，但需要额外的水过滤系统，并会增加维护成本。

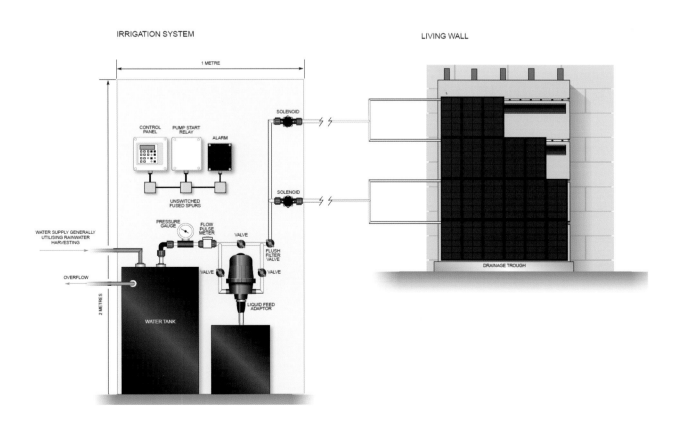

IRRIGATION SYSTEM   LIVING WALL

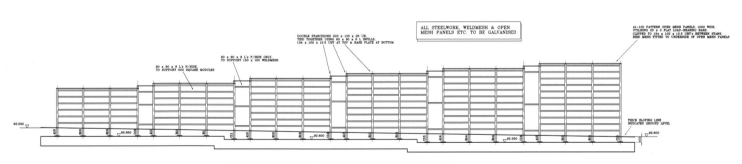

ELEVATION ON GREEN WALL, SHOWING STRUCTURAL STEELWORK ONLY
(VIEWED FROM GREEN WALL PLANTER SIDE, FLATTENED)

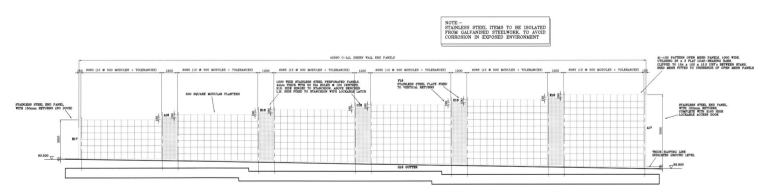

ELEVATION ON GREEN WALL, SHOWING PLANTING BOXES & STAINLESS PANELS ONLY
(VIEWED FROM GREEN WALL PLANTER SIDE, FLATTENED)

Location
Maputo, Mozambique

Maputo Commercial Building
马普托商业大厦

Landscape Architect
Leon Kluge

Building Architect
CNBV architects

Client
Mr. and Mrs. Loreiro

Photographer
Sven Musica

Situated in the heart of Maputo, the Project was the first of its kind in Africa. The building situated in the main road Samora Machel drive in Maputo was to be the home of a bank, a couple of shops and condos on the higher levels.

The green facade of the building faced west, which made it difficult in plant selection because of the mere fact of the severe afternoon African sun, that could easily reach 40 degrees Celsius in summer, and would quickly kill off plants that was not hardy enough to handle it.

The final plant selection consisted of tropical plants such as a wide range of Bromeliads, Alternanthera's, Hemigraphis, Ophiopogon Vittatus, Tillandsias, anthuruims and many more.

The client's brief was to create a modern tapestry of foliage colour, that would keep the modern planting pattern highly visible from the main road with the least bit of maintenance throughout the year.

All plants had to be imported from neighbouring countries as Mozambique did not have the Nurseries to supply the plants needed.

这个设计位于马普托市中心，在非洲算是第一个。该建筑坐落于马普托市主干道萨莫拉·马谢尔快车道，建筑中有一家银行、几家商店，高层是公寓。

建筑物的绿色立面朝向西侧，这给植物选择带来了难度，唯一的一个原因就是，在非洲太阳光太强烈，夏季午后的温度很容易就达到40℃，这会很快杀死那些生命力不太顽强的植物。

最后选择的植被包括一些热带植物，比如，大范围的凤梨科植物、虾钳菜属植物、半柱花属植物、沿阶草属植物、铁兰属植物、花烛属植物，还有很多其他类的植物。

客户要求是要打造一个现代化的、由五颜六色的植物所构成的织锦，以便在维护费用最低的前提下使人们全年都能从主干道上看到这个现代化的种植图案。

所有的植物都必须从邻国进口，因为莫桑比克没有能供应这些植物的苗圃。

Location
Beirut, Lebanon

Stay

Stay 餐厅

Landscape Architect	Client	Area
Green Studios	SOLIDERE	150 m²

Stay Restaurant is the second green wall installation part of the same project. This time Alain Moatti's concept is to have a wavy effect of greenery on the 8m facade of the restaurant's building. The flexibility of our system allows us to achieve this effect of course with the right selection of plants.

As Beirut is known to be a francophone city by excellence and thus strongly rooted with French culture, Sweet tea is managed by French Michelin star chef Yannick Alleno. This somehow consolidates this organic relationship between both countries' culinary cultures and reinforces Beirut's role as the westernized facet of the Arab world and a city which has always influenced its neighbors with state of the art interventions.

The challenges for this installation are the fact that it has two different systems synchronized at the same time, one wtih an open cycle, and the other a closed cycle where the water has to be recollected, recycled and pumped back into the system. Another important challenge is to have a double green wall with a door that opens, to allow access to the technical room at the back.

Green Studios' team is a mix of designers and engineers; this duality in the firm's identity plays an imminent role in the production of both artistic and technologically advanced solutions in the field of Landscape. The additional interesting part of the story is the Mediterranean climate which adds new choices of plants that one can observe in vertical installations and this experience is definitely exciting for any professional in the field of Landscape.

These installations have earned the firm the 2011 MIT award for its scientific findings so when you are traveling to Beirut, remember to look for its green walls, and the experience is worth the visit.

　　Stay 餐厅是同一项目中第二个绿墙安装部分。这次阿兰·莫阿提的理念是在餐厅建筑物 8 m 宽的立面上，用绿色植物打造一个波浪形效果。我们系统的灵活性使我们可以通过正确选择植物达到这个效果。

　　众所周知，贝鲁特是一个讲法语的优秀城市，因此法国文化根深蒂固。甜茶餐厅是由法国米其林星级厨师亚尼克·阿雷诺所经营的。这在一定程度上巩固了国家之间烹饪文化的有机联系，并且强化了贝鲁特的角色——它作为被西方化的阿拉伯世界的一面，时常以其当前发展状况影响着周边地区。

　　这次安装的挑战是它有两个不同的系统需要同步运行。另一个系统有开口循环，一个有闭合循环，在这里，水被重新收集，经过回收利用，再用泵送进系统。另一个重要的挑战是，要建造一个双层的绿墙，并且附带一个可以打开的门，从此门能够进入后部的技术室。

　　绿色工作室团队包括了设计师和工程师，这个双重性，在艺术和技术方面对景观领域先进解决方案的产生，起着迫切的作用。另外，有意思的是，地中海气候增加了垂直绿墙上植物的可选择性。对景观领域的任何专业人员来说，这经验无疑都是令人兴奋的。

　　这些安装设计已经使公司赢得了 2011 年由麻省理工学院颁发的科技成果奖，因此当你在贝鲁特旅游时，请记着去寻找它的绿墙，这值得参观。

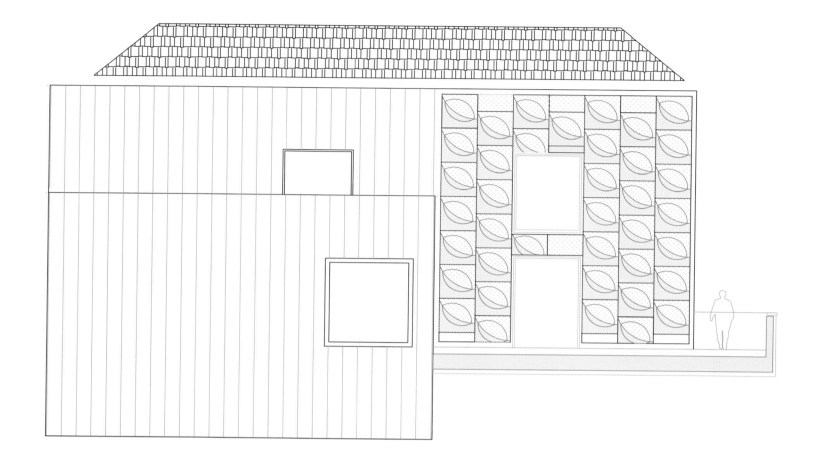

Location
London, UK

Westfield Living Wall
韦斯特菲尔德购物中心垂直绿墙设计

Landscape Architect
AECOM

Consultant
Fountain Workshop

Lighting
Spiers and Major

Client
Westfield Group

Photographer
AECOM

Among the landscape highlights at the high profile retail centre Westfield London, is the 170-metre long living wall. At the western entrance to the 17-hectare scheme, the four-metre high structure is a visually striking and ever-changing landmark creating the backdrop to an upbeat and inviting avenue lined with restaurants, cafes and bars. Standing between the centre and local homes, the north-facing living wall is planted predominantly with native woodland plants, primarily ferns and is incorporated into a massive contemporary-style water feature. In addition to being visually outstanding, the living wall contributes to the environment in numerous ways by helping to filter the air, creating a giant slice of new wildlife habitat in the middle of this retail complex and looking beautiful.

In addition to the wall, the distinctive landscape and public realm design creates a grand entrance, attractive and accessible pedestrian streets and vibrant interior and exterior public spaces. The streetscape incorporates a powerful graphic device of flowing lines in black and silver grey granite paving which draws shoppers into the space from the new underground, overground rail and bus stations completed by the developer.

The generous public realm includes seating, planters, street furniture, trees and ornamental planting. The scheme places great importance on integrating the new development into its urban context with landscape playing a key role in upgrading the local environment and providing screening and buffers between the development and existing houses and transport interchanges.

Concept Illustration

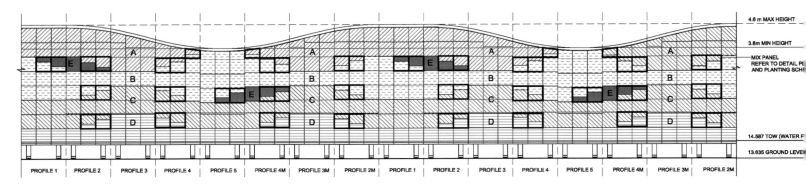

D01 PLANTING PLAN #1- TYPICAL LONGITUDINAL ELEVATION
1:50 @ A0

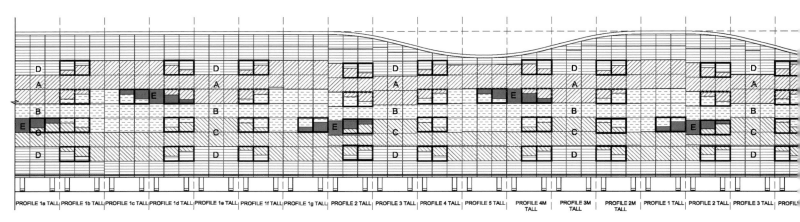

D02 PLANTING PLAN #3 (WEST END)- LONGITUDINAL ELEVATION
1:50 @ A0

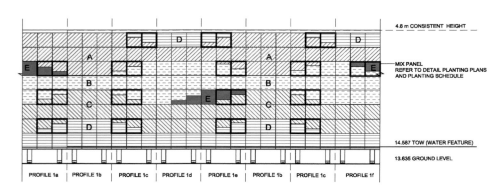

D03 PLANTING PLAN #2 (EAST END)- LONGITUDINAL ELEVATION
1:50 @ A0

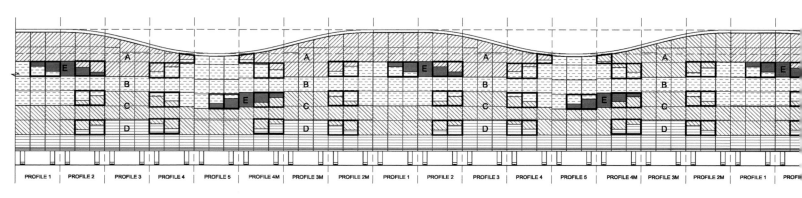

D04 PLANTING PLAN #4 (SHORT WALL)- LONGITUDINAL ELEVATION
1:50 @ A0

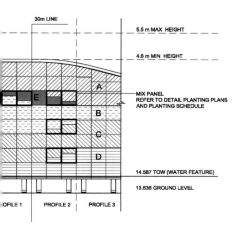

在伦敦韦斯特菲尔德购物中心引人注目的零售商业区中，170 m 长的垂直绿墙成为一道亮丽的风景线。在 17 公顷建筑格局的西入口，4 m 高的绿墙结构给人以视觉上的冲击，大街沿途布满餐厅、咖啡馆和酒吧，不断变化的地标为这个繁荣、诱人的大街提供了背景。位于购物中心与附近区域之间，面朝北的垂直绿墙主要种植的是一些当地的林地植物（主要是蕨类），并融入了非常流行的庭园水景。除了突出的视觉效果之外，垂直绿墙还在很多方面对环境有益：帮助过滤空气，为野生动植物在零售综合中心打造一大片新的栖息地，另外，还起到美化的效果。

除了绿墙之外，有特色的景观和公共领域设计建立了一个壮观的入口、美观便利的人行街道和富有生气的室内外公共空间。街道景观中有一个由黑色和银灰色花岗岩铺成的走道，该走道独具特色，呈流线型，吸引购物者从开发商新建的地下、地上轨道和公交车站进入该空间。

宽敞的公共领域内包括座椅、花架、街道家具、树木和观赏植物。该方案很注重将新建筑与城市文脉融合在一起，该景观在多方面起到了重要作用，如提升当地环境，在新建筑与现存房屋和交通交汇处之间提供遮蔽和缓冲。

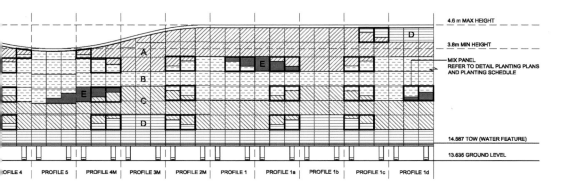

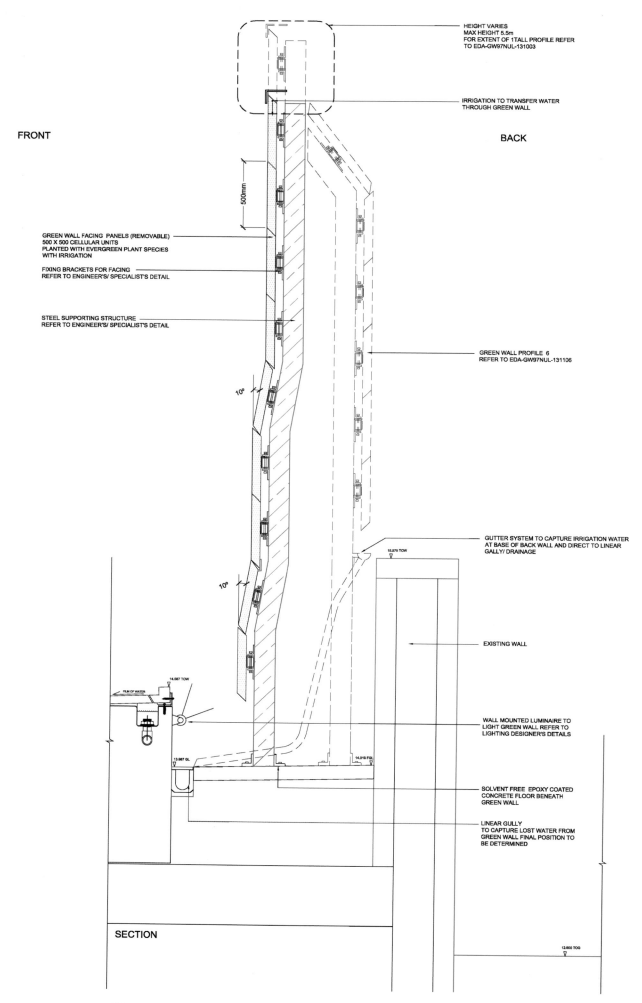

winter

spring

summer

autumn

ferns helleborus wood anemones violets snow drops liriope hosta Bergenia

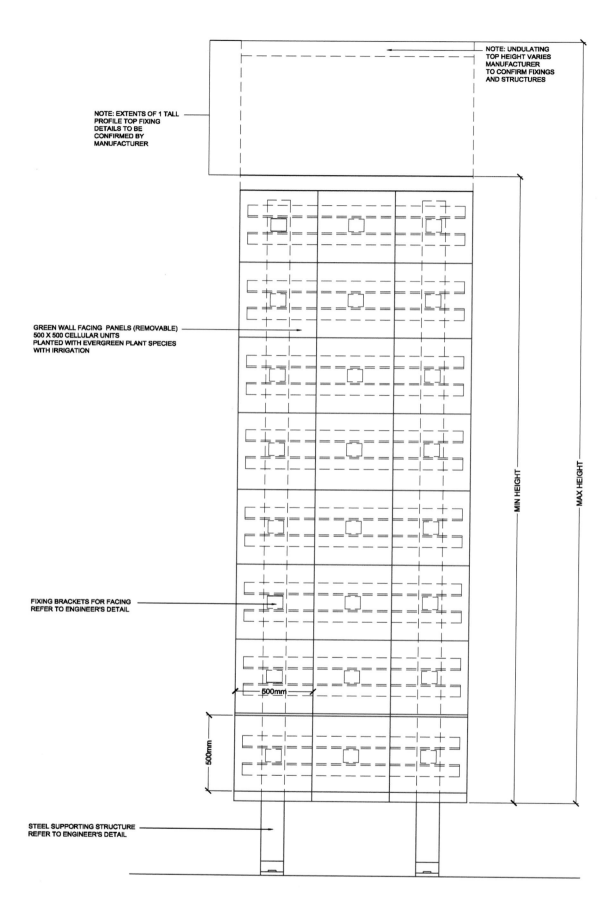

D02 ELEVATION: GREEN WALL PROFILE 1, 1 TALL
SCALE: 1:10 @A0

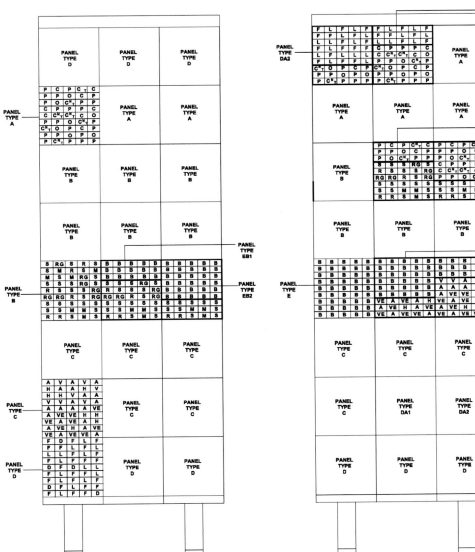

PROFILE 1d

PROFILE 1e

PROFILE 1f

D02 PLANTING PLAN: PROFILE 1d, 1e, 1f
SCALE: 1:10 @ A0

Location
Lima, Peru

Zentro Office Building and Commercial
Zentro 商用办公大楼

Landscape Architect
Oscar Gonzalez Moix

Firm
Gonzalez Moix Arquitectura

Client
Antonhy Isles Infanta

Photographer
Juan Solano Ojasi

We have to solve a typical project, a design center, where creativity and art should be blended and be offered to the users. The shops should have the same condition, without creating differences between the front and back. After solving the first floor with two locals, we make in the second and third floor 8 offices or locals arranged beside a yard with domestic scale.

This courtyard is limited by a green wall of recycled wood and plants that spring from it. The artist Veronica Crousse is called to design this large wall sculpture made from recycled materials with living vegetation in the gaps left by the position and movement of the wood inserts. It generates a very special life to a dividing wall which coexists with the everyday creative work.

The courtyard stars the project and gives life to these locals. This resolves an equitable and warm condition that coexists with the creative environment of the center. Thus, between the 8 offices and the vertical garden the terrace is born, an open space that acts as entrance court for workers and visitors of Zentro.

The language of the project achieves harmony between contemporary materials that give strength to the construction. Details in exposed concrete, recycled wood and stainless steel define the small but important spaces in the project. One example is the staircase in the office atrium, being open and playing with this space, therefore it mixes the rusticity of the wood planks with the subtlety of the steel to make an essential piece of the project.

In short, Zentro represents not only a project well-studied that inserts into an urban environment, but also is itself a set of spaces and sensations that arise as one delves deeper into the proposed architecture.

LEYENDA

1 Estacionamiento
2 Depósito
3 Espacio comercial
4 Plaza
5 Oficina
6 Kitchenet

YARD SECTION   ELEVATION   GREEN WALL SECTION

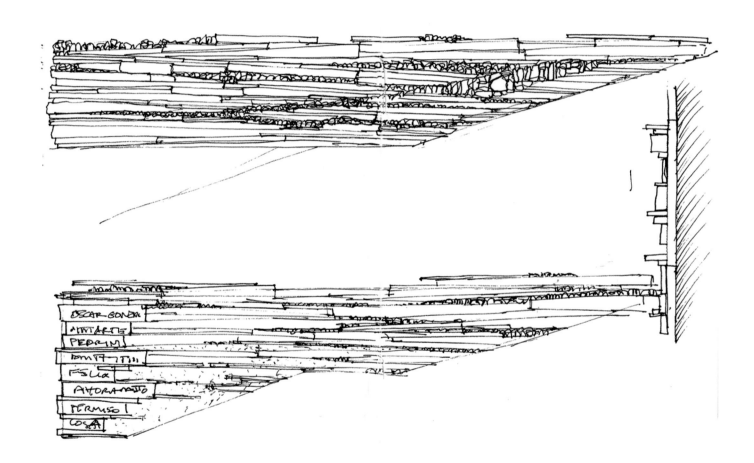

本案为典型项目，需为用户提供一个将创造力与艺术融为一体的设计中心。商店的设计亦是如此，前面和后面毫无二致。我们首先解决了一层的两个区域，之后解决了二层和三层的 8 间办公室，以及内部院落旁边的几个区域。

本案中的院子里有一面绿墙，绿墙由可再生树木和其中生长的植物构成。艺术家 Veronica Crousse 受托设计这个大型的墙体雕塑，雕塑由可循环材料构成，木材间隙生长着绿色植物使这面隔断出日常创造性工作空间的隔断墙呈现出独特的生趣。

本案中庭院设计是项目的点睛之笔，给每一个区域带来勃勃生机，为中心的创意环境创造了一个公平温暖的氛围。因此，8 间办公室和垂直花园之间形成了一处平台，一个开放的空间，可以作为工人和访客进入 Zentro 的入口。

本案在当代材料之间取得了平衡，这些材料是建筑的特色。本案中的空间虽然小，但是很重要，空间充满了细节，如暴露的水泥、可再生木材和不锈钢。其中一个例子就是办公楼中庭的楼梯，楼梯是敞开的，与空间浑然一体，它将木板的纯朴和钢的细腻融合在一起，成为项目的重要组成部分。

简而言之，Zentro 项目经过深思熟虑，与城市环境相得益彰，同时，当人们深入其中，他们可以感受到项目提供的一系列空间和灵感。

Angie Fowler Adolescent & Young Adult Cancer Institute

Angie Fowler 青少年及年轻人癌症研究所项目

Location
Cleveland, Ohio, USA

Landscape Architect
Virginia Burt, CSLA, ASLA, Visionscapes Landscape Architects Inc.

Project Architect
Stanley Beaman & Sears

Client
University Hospitals Rainbow Babies & Children's Hospital

Photographer
Hanson PhotoGraphic

The Angie Fowler Adolescent and Young Adult Cancer Institute project leadership and primary donor sought out and championed design solutions for the complete 8th and 9th floors' renovation that were thoroughly engaged in creating a light-filled space of healing for the patients and an environment fostering community and togetherness.

To that end, the 8th floor outpatient architecture has been developed as a journey of light and healing. A 60 feet long, custom illuminated welcome wall guides patients from the lobby to treatment wings where decentralized care team stations allow direct communication between staff and patients. Patients are able to personally adjust the color of the treatment spaces through color changing light fixtures incorporated into the ceilings. Magnetic glass walls allow for personal messages as well as staff communication. Preserved views and windows are impactful throughout space, but if quiet reflection is desired, frosted glass doors can be closed to create a more private environment. Patient care extends beyond the clinical world into the realm of the emotional, experiential, personal and possible.

From waiting areas to pediatric activity rooms to the teen lounge and the treatment rooms themselves, the entire project is designed to embrace everyone who enters with warming light-filled spaces and to provide places to help the patients heal. Three key client principles: participation, collaboration, and sharing information, guide the planning and decisions along the path to completion. Zones of interaction between patients and staff, patients and their families and friends, patients and other patients are found throughout the project. Decentralized care team stations and custom glass allow informal communication between patients, nurses, and doctors.

The jewel in the crown is the completely reimagined 9th floor. What was once a bare, exposed roof has been transformed into a sky lobby and an extensive outdoor rooftop garden, completed with a green wall, whimsical sculptures, intimate planters, green hills, shady trees, inviting spaces, and a soaring multicolored glass canopy. Everything is meticulously coordinated to provide the much-needed moment of respite from the healing journey that begins below.

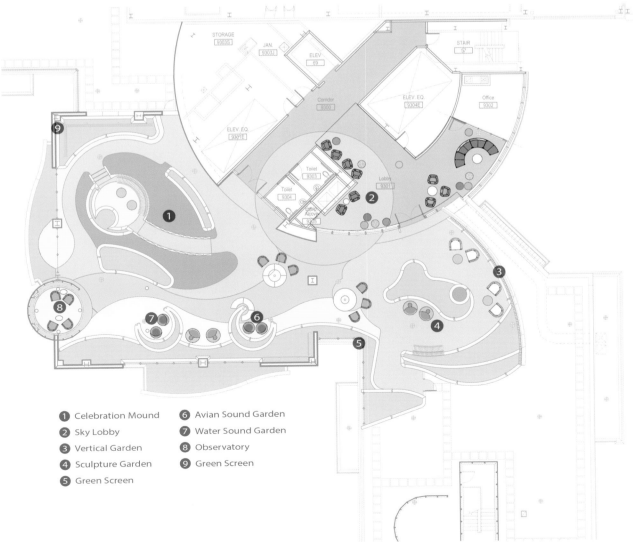

1. Celebration Mound
2. Sky Lobby
3. Vertical Garden
4. Sculpture Garden
5. Green Screen
6. Avian Sound Garden
7. Water Sound Garden
8. Observatory
9. Green Screen

Angie Fowler青少年及年轻人癌症项目的领导和主要捐款人旨在寻找一个对第8~9层楼进行整体更新的解决方案。他们力争为病人创建一个采光充足的治疗空间，一个培育社区感和归属感的环境。

为此，本案对第8层楼的门诊部进行了改造，使其成为一场光照充裕的治疗之旅。门诊部有一处18m长、墙上配有照明的定制欢迎墙，引导病人从大堂前往分散在翼楼各处的治疗室。病人可以在治疗室与医生直接交流。病人还可以通过天花板内安装的光线调节装置调节治疗室的颜色。磁化玻璃墙使个人交流和员工沟通成为可能。空间处处保留有窗户，如果人们需要安静思考，可以关上磨砂玻璃门来创造一个更隐秘的环境。对病人的护理除了临床以外，还包括情感、体验以及私人的照料。

从休息区到儿科活动室，到青少年休息室和治疗室，整个项目设计得让人感觉温馨洋溢，走入充满温暖光照的房间，任何人都会感觉宾至如归，为病人们提供了一个治疗的环境。设计体现了三个关键的客户原则：参与、合作和信息共享。这些原则将指导设计师沿着既定计划完成规划和决策。本案设置了交流区，病人可与医护人员、家属及朋友，或与其他病人进行交流。分散的治疗室和定制的玻璃使病人、护士和医生之间可以进行非正式的交流。

第9层楼是本案的点睛之作。本案将曾是光秃秃的、完全暴露在外的屋顶改造为一座开阔的屋顶花园，有绿墙、异想天开的雕塑、植物、绿山、树荫，整个空间格外诱人。顶上是多彩的玻璃天棚。所有元素都经过精心设计并相映成趣，为治疗过程提供了一个很必要的休息空间。

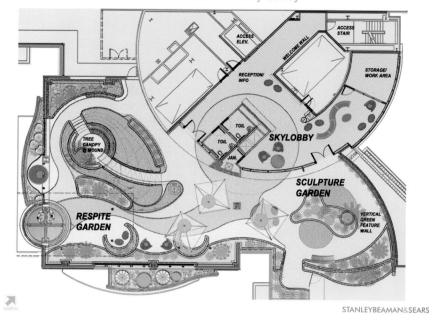

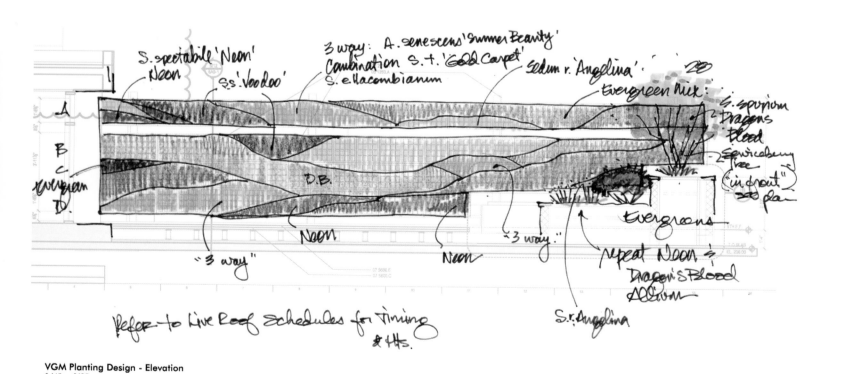

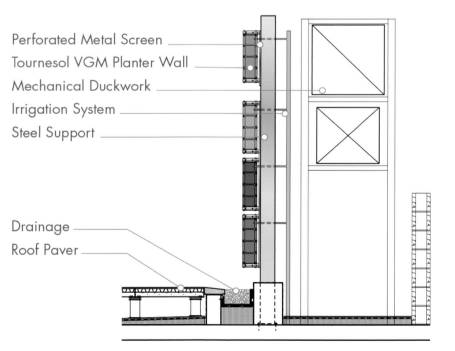

- Perforated Metal Screen
- Tournesol VGM Planter Wall
- Mechanical Duckwork
- Irrigation System
- Steel Support

- Drainage
- Roof Paver

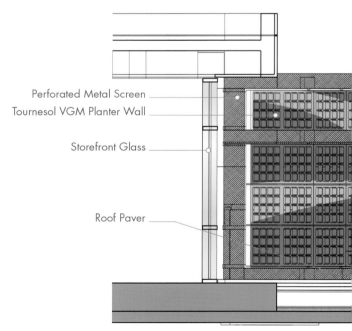

- Perforated Metal Screen
- Tournesol VGM Planter Wall

- Storefront Glass

- Roof Paver

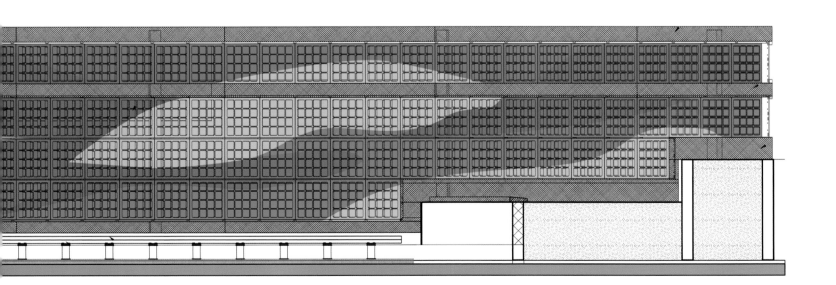

Location
Beirut, Lebanon

BB House
BB 大楼项目

Landscape Architect
Green Studios

Client
BB House

Area
140 m²

Green Studios handles the green spaces in the BB House building. The scope includes both designing and building for the ground floor/entrance and the roof vertical garden spaces.

The ground floor, designed in a chess-like grid, is the entrance to the building, and a link between the outdoor sidewalk and the indoor art gallery.

The garden is designed according to the spirit of the art pieces and the interior design of the gallery. The use of stainless steel edges flows from the outside to the interior of the building.

The green walls on the other hand, located on the roof floor of the building, surround the company's glass framed meeting room. The roof wrapped with the green walls is used as the outdoor "garden" space. The building is located in Hamra which is the very Heart of Beirut, and is overcrowded by high-rise buildings; so the need for the vertical walls is very high. The concept behind the green walls aims at overwhelming the space with green, in contrast to its very rigid surrounding. The planting layout and selection of lush varieties portray a natural feel.

绿色工作室负责处理 BB 大楼内的绿色空间，包括设计和建造一层／入口以及屋顶垂直花园空间。

一层设计成棋盘状的网格形，是大楼的入口，连接了一条户外人行道和室内艺术画廊。

花园是根据艺术品的精神和画廊室内设计来设计的。整个建筑由内到外均使用了不锈钢包边。

另一方面，位于大楼顶层的绿墙围绕公司的玻璃框架会议室而建。屋顶包裹在绿墙内，可用来作为室外的"花园"空间。本案大楼位于贝鲁特市中心的 Hamra。这里高楼林立，因此急需垂直绿化墙。本案旨在用绿色征服空间，用绿色与周围的冷峻环境形成对比。郁郁葱葱的植物传递了一种大自然的感觉。

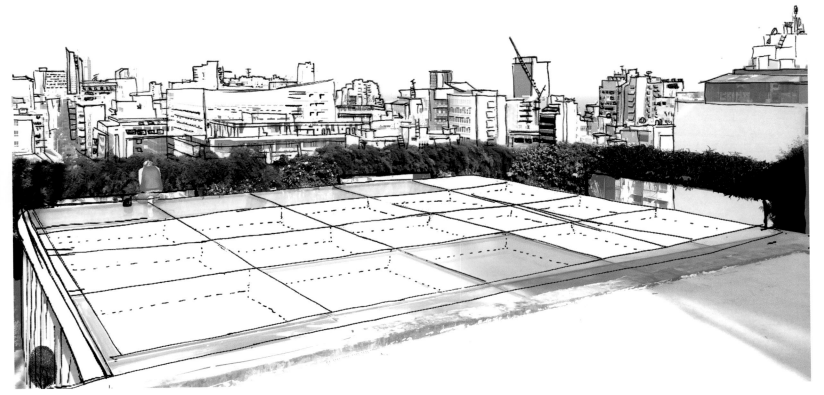

Firma Casa

Firma Casa 商店

Landscape Architect
SuperLimão Studio, Studio Campana

The Firma Casa project started to be developed in November 2008, when Sonia Diniz Bernardini, owner of Firma Casa, decided to renew her store, established in 1994. She invited the Studio Campana to make the project and they decided together to invite SuperLimão Studio, a young architecture and design studio, to make the project and develop a lot of ideas.

The project consists of a two-floor building with 500 square meters divided into a gallery, a retail store and, in the second floor, the offices. All of the steel structure, air conditioning ducts, and a grid of electric rails are shown in the ceiling. The beams can be used with industrial magnets to hang pieces, and pallets shelving can support different pieces with different dimensions. SuperLimão Studio seeks for flexibility to develop the project which could be used for a lot of different exhibitions.

A three-piece front door allows pieces of big dimension to enter the gallery and the whole concrete floor can support heavy objects, sculptures, etc. In the outdoor area the Elastopave® is used to give the floor the capacity to drain rainwater.

To cover the whole façade, Fernando and Humberto Campana suggests a green wall with Espada-de-São-Jorge(Sansevieria Trifasciata), a plant of African origin and very diffused in Brazilian popular culture. In front of this challenge SuperLimão Studio designs and develops a bent aluminum vase with an origami form to support the plants. There are 3500 vases with 9000 seedlings of Espada-de-São-Jorge.

225

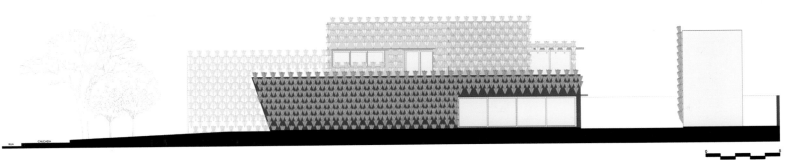

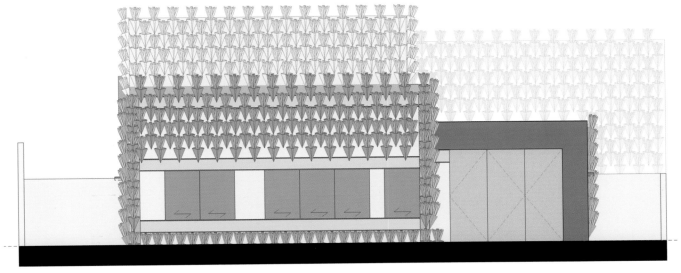

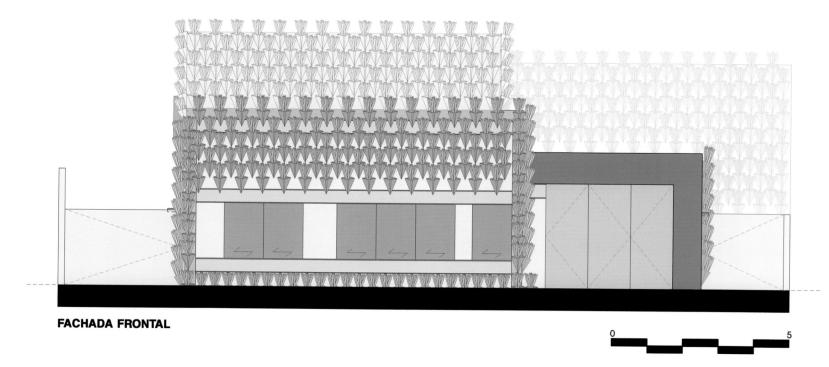

FACHADA FRONTAL

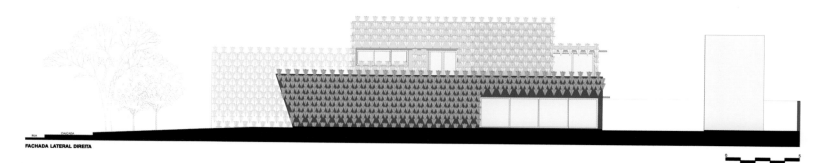

FACHADA LATERAL DIREITA

Firma Casa 商店这一项目开始于 2008 年 11 月,当时 Firma Casa 的主人 Sonia Diniz Bernardini 决定翻新这个始建于 1994 年的商店。她邀请了 Campana 工作室来做这一项目,而后他们共同决定邀请了 SuperLimão ——一个年轻的建筑和设计工作室,一起来做这个项目,并形成了很多的创意。

这个 500 m^2 的建筑分为两层:一层是一个画廊和零售店;二层是办公室。建筑采用钢制结构,安装有空气调节管道,屋顶上可以看到网格式的电动轨道。带有磁性的横梁和平板支架可以悬挂或放置大大小小的各种物品。SuperLimão 工作室灵活地开展这个项目,使该设计可以被用于许多不同的展览会。

前门由三部分组成,能够使大尺寸的部件进入画廊,整个混凝土地板可以支撑重量大的物体和雕像等。在室外区域,为了使地面有足够的排水能力,半透水路面构造被采用。

为了覆盖整个建筑物立面,Fernando Campana 和 Humberto Campana 建议用金边虎尾兰建造绿墙,金边虎尾兰源于非洲,在巴西大众文化中很常见。面对这个挑战,SuperLimão 工作室设计和开发了一种弯曲的、带有折纸形状的铝花瓶来支撑植物。整个设计共有 3500 个花瓶,9000 株虎尾兰小苗。

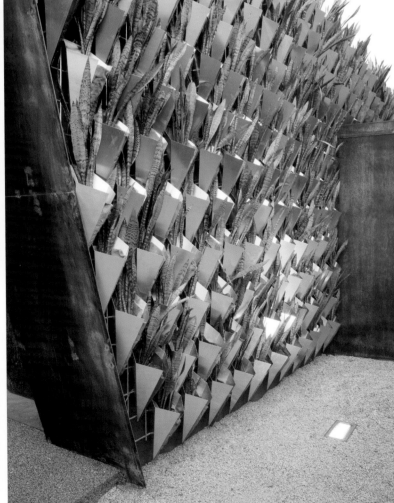

Location
Puerto Vallarta, Mexico

巴亚尔塔住宅区

Landscape Architect
Ezequiel Farca

EZEQUIELFARCA architecture & design

Area
3,000 m²

Photographer
Jaime Navarro, Roland Halbe

Located in the Marina of Puerto Vallarta in Mexico, the House takes advantage of a beautiful view on the Pacific bay of Vallarta.

The house was conceived to allow family and guests to enjoy the peaceful surroundings. It has all the amenities allowing a great experience during a journey in Vallarta: a fitness center, a home theater, 2 Jacuzzis, a terrace with a swimming pool, a fire pit, 8 bedrooms, and so on. Its main focus was to create multifunctional spaces that could open to the exterior. Every room has floor to ceiling windows that open up to private terraces; Vallarta's unique climate permitted us to design these kind of areas. The main attraction is the living and dining room, which is surrounded by curtain glass walls that open and unify with the pool and terrace area, creating a strong experience of bringing the outside in the heart of the house.

The use of natural stones and concrete walls creates fresh spaces that help during hot seasons. It was also important to use materials that brought warmth to each area, which was achieved by creating a kind finish that has the appearance of wood but was made out of concrete panels. This finish was used mainly on the façade, but was also seen in some of the important indoor spaces.

The objective of installing green walls and rooftops was to insulate the house from heat, reducing the use of air conditioning. They also help integrate the house with its landscape. Because the house is located below a multistory residential building, it was important not to interfere with their ocean views, and give the rooftop a nice finish.

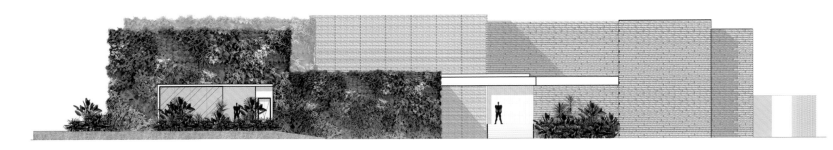

坐落于墨西哥巴亚尔塔港的码头，该住宅借助了巴亚尔塔港太平湾的美景。

这个房子的设计是为了让家庭成员和客人们享受安静的环境。它拥有齐全的设施：一个健身中心、一个家庭影院、两个热水浴缸、一个带游泳池的平台、一个火坑、8个卧室等，使人们在巴亚尔塔港旅行期间可以获得美妙的体验。它着眼于创造可以对外部开放的多功能空间。每一间房间都有海景落地窗，可以通向私人阳台；巴亚尔塔独特的气候条件使我们可以对这种区域如此设计。主要的吸引人之处是由玻璃幕墙所包围的起居室和餐厅，这些玻璃幕墙可以打开，与游泳池和阳台区域融为一体，从而产生一种将外部世界带入房屋中来的强烈体验。

使用天然石材和混凝土墙，使空间清爽，这在炎热的季节作用尤其重大。使用保暖材料对每个区域来说，也很重要，这一点通过创造一种独一无二的抛光来实现，该抛光有木制的外表，但由混凝土板材制成。它主要用在建筑的立面上，也可以在一些重要的室内空间见到。

安装绿墙和绿色屋顶的目的是使房间隔热，以减少空调的使用，也利于房屋与景观的融合。由于房屋坐落于多层住宅建筑的下面，所以它不能跟海景相抵触，同时还得给房顶一个漂亮的抛光，这很重要。

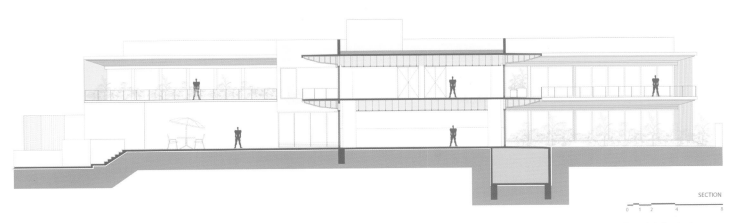

section_Vallarta house

Location
Brussels, Belgium

Restructuration of the Brussels Regional Council

布鲁塞尔区政府重建工程

Landscape Architect
P. LEBLANC & P. SPEHAR

Building Architect
Art & Build Architects

Pursuing its installation on the site of the old Governor's Palace, the Brussels Regional Council recently entrusted to the architectural consultancy, Art & Build, the restructuring of the interior of the island formed by the rue du Chêne and the rue du Lombard. The operation forms part of the general policy of slum clearance undertaken by the public authorities of the Region of the City which are located in their respective territories. After having "oxygenated" the fabric of the buildings which is of excessive thickness by freeing it from insalubrious chancres, the architects made use of the differences of level of the two arteries and projected a set of installations beneath the hanging garden (a polyvalent hall and annexes for the use of the inhabitants of Brussels and their elected representatives). This new tool will create an essential link between the different components of the installations and makes possible an unobstructed and fluid movement between the semicircular building, the parliamentary offices, the reception rooms and the administrative departments. The setting of this 'city garden' designed by the architects is based on the central theme of an agora, a place of democracy par excellence. A wide wooden platform makes possible informal exchanges amid aromatic plants, which reflect changing seasons which are noticeable in the heart of the city. The rhythm and sequence of the design affords areas of calm, which are propitious to meditation and reflection, whilst protecting the work areas behind the screens of verdure. The design of a water fountain will be entrusted to an artist of repute and will complete the consistency of the intervention.

The layout of the garden will be based on a dense and diversified environment. The object is that of showing to their best advantage those buildings possessing a special character, such as the Palace which houses classified rooms, as well as the new semicircle, or again a building dating from the XVIIth century.

为了在老总督府的原址上进行安装，布鲁塞尔区政府最近委托建筑咨询公司——艺术和建筑，进行由 Chêne 和 Lombard 两条街围成的岛区的内部重建。该工作是各自管辖区内政府当局所承担的棚户区改造工程的一部分。建筑师们给建筑结构"充氧"，剔除建筑中的顽疾，利用两个要道的水平差，在空中花园下方设计出一组装置（一个多功能大厅和附属建筑物，供布鲁塞尔的居民和民意代表们使用）。这个新的工具将在不同的安装部件之间建立一个基本的联系，使半圆形建筑、议会办公室、接待室和行政部门之间能够畅通无阻地运行。"城市花园"的布置是由建筑师根据"城市广场——最民主的地方"这个中心主题而设计的。一个大的木制平台使人们可以在芳香植物中进行非正式的交流，这些植物可以反映出市中心明显的季节变化。设计的节奏和顺序化，为人们提供了适合冥思和沉思的安静之地，同时，绿色植被的遮蔽保护了工作区域。喷泉的设计将委托一名有名望的艺术家，来实现设计的一致性。

花园的布局将以浓密、多样化的环境为基础。目标是最好地展现那些具有特殊品质的建筑物，比如，有着各式房间的宫殿和新的半圆形建筑，或者起源于 17 世纪的建筑。

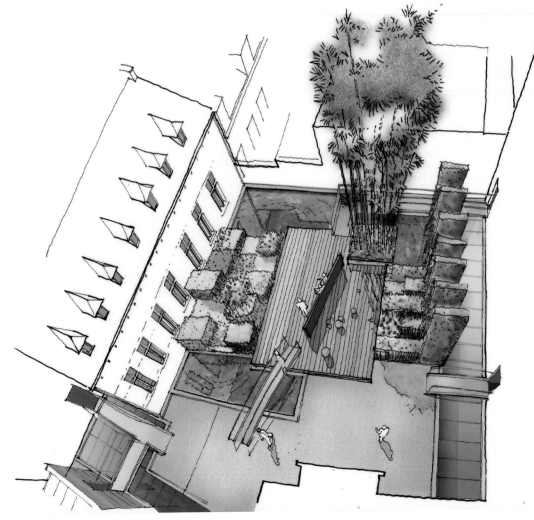

Situation existante

Situation projetée

Location
Barcelona, Spain

Green Side-wall
绿色侧墙

Landscape Architect
Capella Garcia Arquitectura S.L.P.

Building Architect
Juli Capella, Miquel García

Area
288 m²

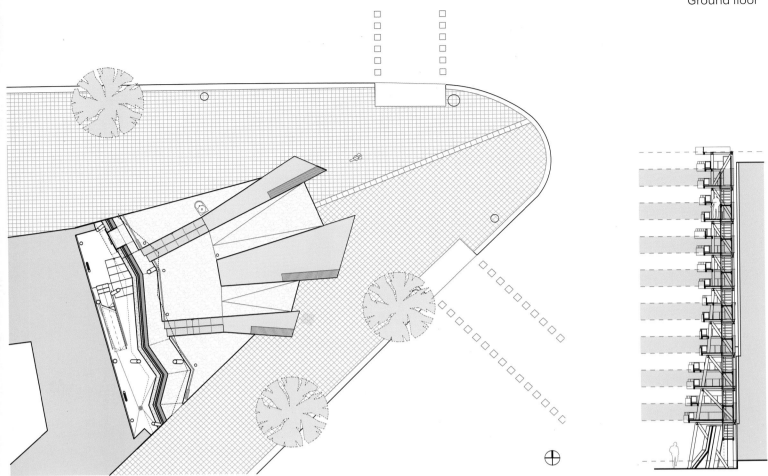

Ground floor

The Green Side-wall consists of a free-standing structure containing plants that form a protective mass of vegetation against a facade in Barcelona, thus creating a vertical garden.

From street level, the structure becomes gradually narrower as it rises, to a height of 21 metres. It consists of a ground-floor level and eight upper levels accessible by way of interior steps. From the first to the eighth level, metal platforms are provided on which the flower-trough modules are arranged perimetrally, on two distinct levels. These platforms can be reached, with restricted access, from the ground floor by way of interior steps. This convenience of access is precisely what differentiates this structure from other vertical greenery, maintenance and replanting of which always has to be done from the exterior using elevating platforms, making the process a difficult and expensive one requiring specialised labour.

For the maintenance of the vegetation that forms the facade, account has been taken of all current requirements for cleaning, safety and sustainability (including a pulley system to transport materials). Water consumption is minimised by means of an automatic programmed drop-by-drop irrigation system with controlled drainage and automatic dosing of fertiliser. Nesting boxes are also integrated. In general, the project has been inspired by the concept of "xeriscaping", which advocates the rational use of irrigation water, the planting of local species, ecoefficient design and maintenance criteria.

The free-standing structure is inspired by the form of a tree. The design of each level is different, giving the facade a dynamic appearance. Furthermore, its faceted shape benefits the distribution of the plants, enabling each variety to receive greater or lesser insolation. It forms a great three-dimensional screen, made up of the metallic structure and the flower troughs, which can accommodate different varieties of vegetation and flora. A total surface area of 200 square metres can display all sorts of shades and textures of green, or indeed different gamuts of floral colours. It is therefore a dynamic facade, always alive and always different.

绿色侧墙位于巴塞罗纳，包括一面独立墙结构，其上面的植物在表面形成了具有保护性的大片植被，从而形成一个垂直花园。

从街面来看，随着绿墙逐渐升高（一直达到21 m高），它的结构逐渐变窄。它包括底层和可以从内部楼梯进入的上面8层。从1层到8层，有金属的平台，花槽模块被安排在周边不同的层面上。从底层经由内部楼梯，通过特定的通道到达这些平台。精确地说，进入的便利性是这个结构与其他垂直绿化的不同之处，其他垂直绿化的保养和移植需要从外部使用升降平台来完成，整个过程难度很大，费用很高，并需要专门的人工来完成。

对于立面的植被保养，目前清洁、安全和持续性（包括一个滑轮系统运输材料）都被考虑在内。通过自动滴灌系统程序，将水消耗降到最低，并且控制排水和自动定量施肥。系统中还装有巢箱。总之，这个设计受到"节水型园艺"理念的启发，它提倡理性运用灌溉水源，选择当地植物品种，采用高效节能型设计和保养标准。

独立墙结构受到树形的启发。每一层的设计都不同，立面呈现动态的外貌。而且，多面的形状对植物的分布有益，使每一种类都可以得到更多或更少的日晒。它形成了一个由金属结构和花槽组成的大的三维屏风，可以容纳各种各样的植被和植物群。表面总面积200 m²，可以显示各种各样的绿色植物的阴影、纹理，或者不同的花色色域。因此，立面富有活力和生气，与众不同。

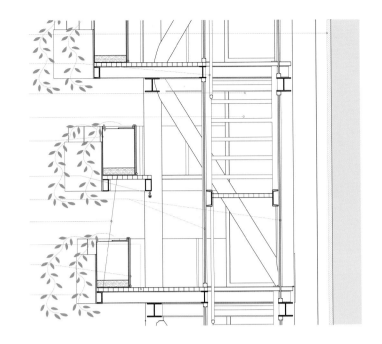
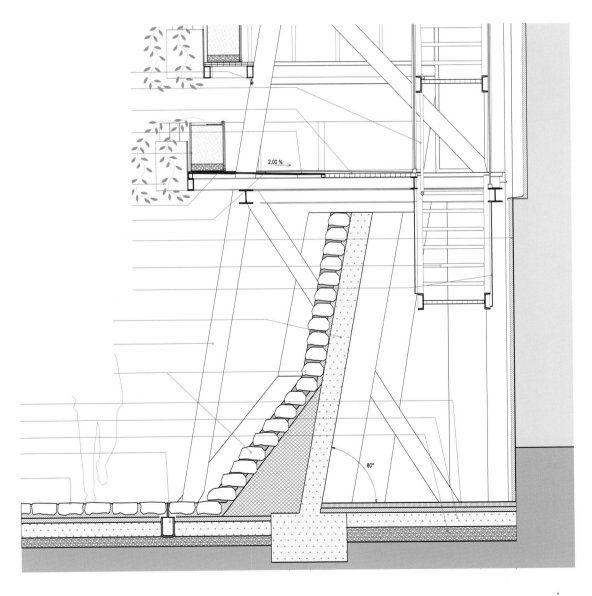

Construction detail

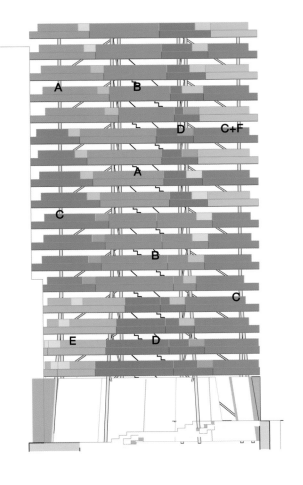

Vegetation scheme

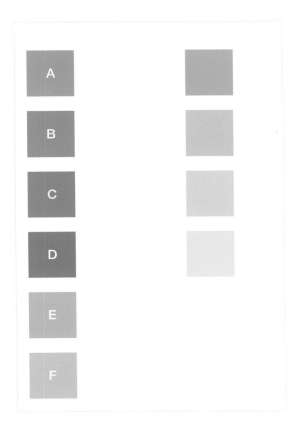

Location
San Diego, California, USA

Fashion Valley Mall
时尚谷购物中心

Landscape Architect
MLA (Mission Landscape Architecture)

As part of a major renovation of Fashion Valley Mall in San Diego, California, the mall owner Simon Properties enlisted the professionals at Mission Landscape Architecture (MLA) to transform an unsightly "back of house" area into a memorable space that would provide a grand focal point at the terminus of the Main Entry drive as well as a waiting area for patrons of an adjacent new high-end restaurant.

To that end, MLA landscape architect Rocco Campanozzi and Jerri Pick design an eight hundred square feet "Living Wall", a vertical garden that backdrops the newly created seating area and has become its own destination. The linear design, based loosely on the keys of a piano, demonstrates the importance of proper plant selection including eventual size, foliage color and texture, water and light requirements in achieving a successful installation. This challenging north facing location requires shade tolerant plants. Mondo Grass and Moneywort make up the background foundation while Spider Plant, Carpet Bugle, Moses-in-a-boat and Coral Bells provide the undulating variations in size, texture and color that make this wall so interesting. Beyond the plant palette, selection of the appropriate planter system, irrigation system, drainage method, type of lighting and installation methods are critical factors in the design criteria. MLA works closely from initial design concept through installation with the other team members: John Willingham from Tournesol Siteworks – planter system manufacturer, Armstrong Nurseries – supplier and growers of planting materials, Jim Mumford of Greenscaped Buildings – irrigation and planter installer, and of course Simon Properties – owner.

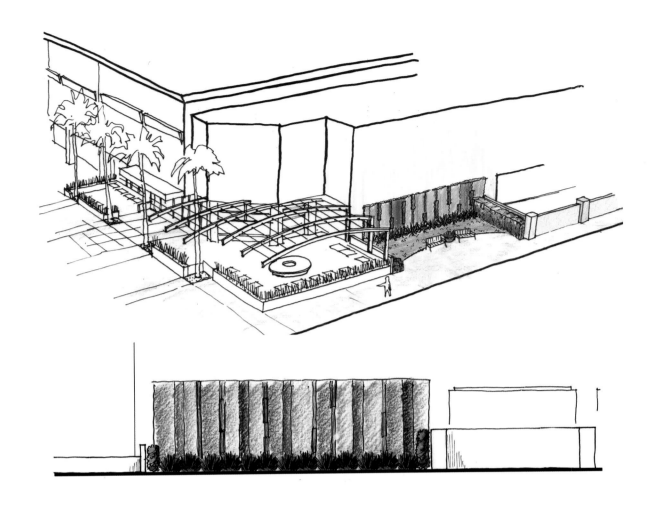

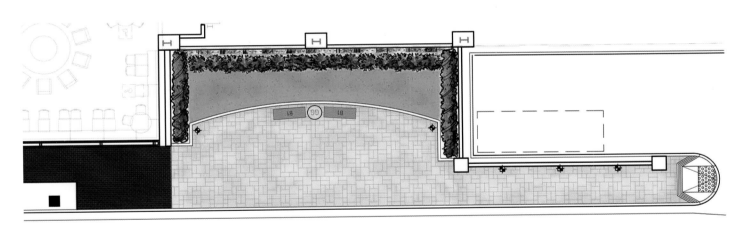

为了对加利福尼亚州圣地亚哥市的时尚谷购物中心项目进行重大革新，商场拥有者西蒙地产委托景观建筑任务组（MLA）的专业人士，拟将这座不起眼的"房子后部"改造成令人难忘的空间，空间将在主入口车道的终点站为人们提供一个宏大的焦点，并为附近一家新的高端餐厅的客户提供等候区。

为此，MLA的景观设计师Rocco Campanozzi和Jerri Pick设计了一座74 m²的"垂直绿化墙"，这座垂直绿化墙可作为新创建的座位区的背景，并且自身也成为目的地。线形设计，以松散的钢琴键的形式，表明适当选择植物是非常重要的，包括植物最终的大小、树叶的颜色和纹理，以及在成功安装时需要的水和光。因为墙体朝北，所以需要选择耐阴的植物。垂直绿化墙以粗草、珍珠菜为背景，此外，吊兰、筋骨草、紫万年青、珊瑚钟等植物的添加使绿墙富于起伏变化，不同的质地和颜色也使这面墙妙趣横生。除了植物的配色，在设计标准中选择适当的种植系统、灌溉系统、排水方法、照明模式和安装方法也至关重要。

从最初的设计理念到安装过程，MLA的设计师与其他团队成员密切合作：来自种植系统生产商Tournesol Siteworks的John Willingham，植物供应和栽培商Armstrong Nurseries，灌溉和花架安装公司Greenscaped Buildings的Jim Mumford，当然还有物业拥有者西蒙地产。

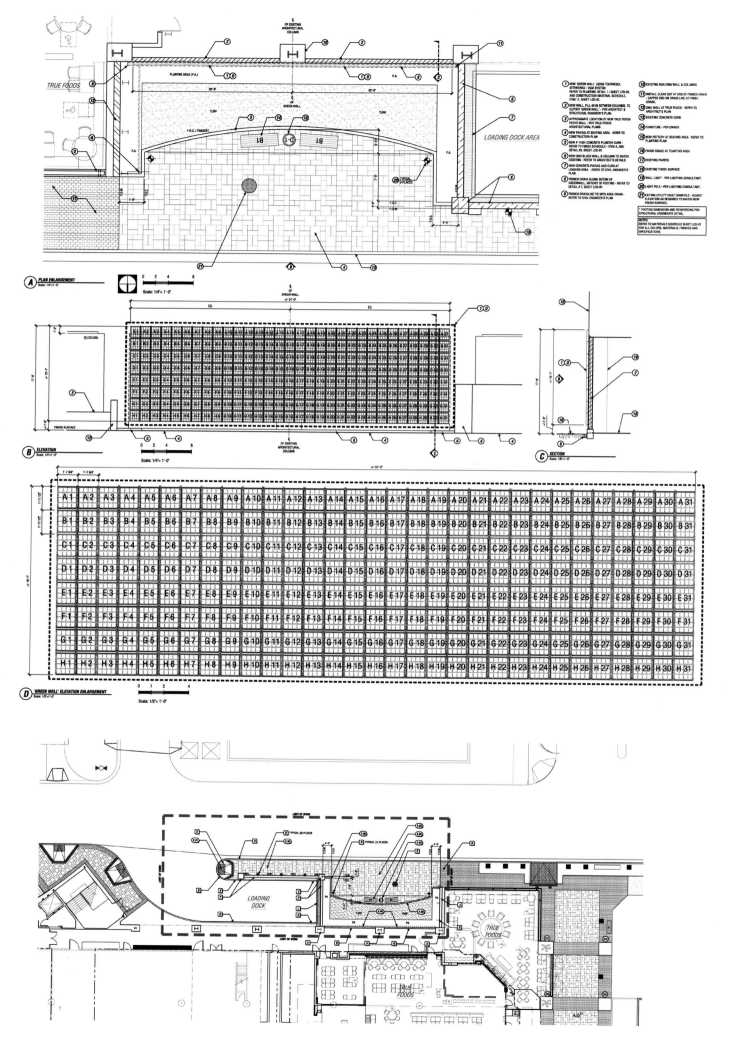

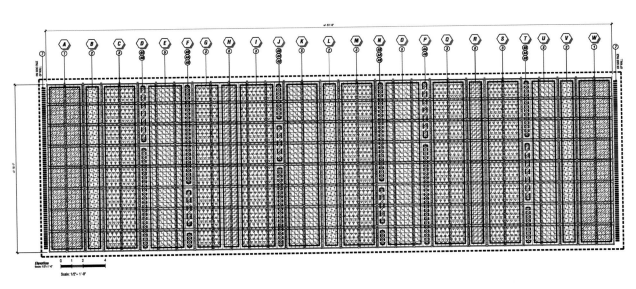

GREEN WALL - PLANTING SCHEDULE

Illura Apartments

Illura 公寓

Location
West Melbourne, Victoria, Australia

Landscape Architect
Tract

Building Architect
Elenberg Fraser Architects

Constructor
Fytogreen Australia

Area
22 m²

Photographer
Fytogreen Australia

This Architecturally designed modern apartment building is located in an older inner city suburb of West Melbourne. The contemporary style, mixed with old suburban buildings, works well with the vertical garden to complete the design. The vertical garden is made up of 4 sections and used as a street view making a powerful impact.

The elevated sections of the garden face northeast hence they are all drought tolerant and sun hardy species.

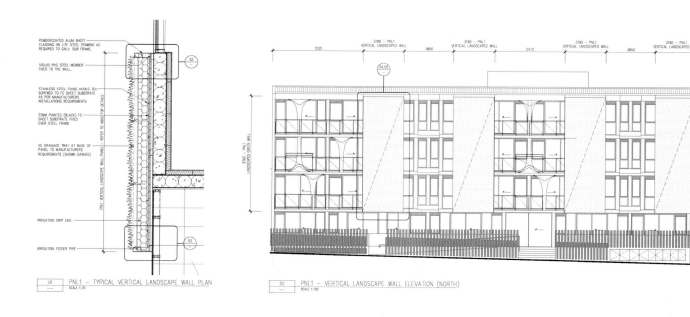
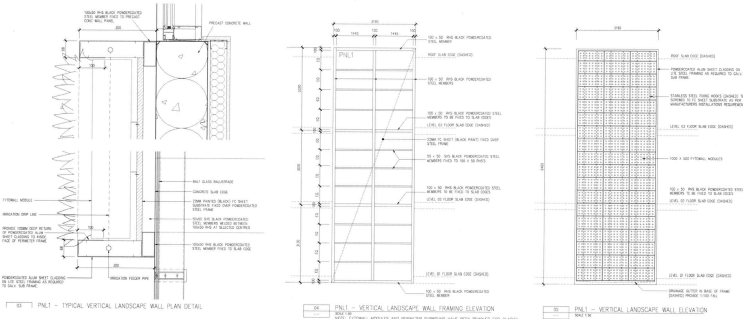
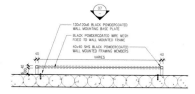
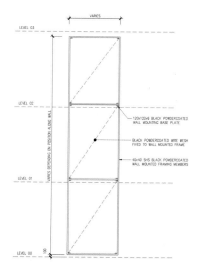

这座体系结构设计的现代公寓位于西墨尔本老城的郊区。该公寓的现代加上垂直花园与旧郊区建筑混合在一起，垂直花园与其相得益彰，使设计更加完善。垂直花园由四部分构成，花园作为街景在视觉上也很有冲击力。

花园抬高的部分朝东北向，因此花园里的植物全部是耐旱和耐阳的品种。

Location
Prague, Czech Republic

Jindrišská 16

Jindrišská 16 项目

Landscape Architect
DAM Architects

Plant/Greenery Design and Realization
Arch Vegetal

Main Engineer
m3m

Area
6,600 m²

Client
IMMOFINANZ Group

Photographer
Filip Slapal

Jindrišská 16 was originally built at the beginning of the 19th century. The new project deals with alteration of the original street building wings and the addition in the courtyard block. The historical building's ground floor consists of rental areas and a café shop/restaurant extended to the atrium. This atrium is established by roofing the former courtyard by a glass roof forming a generous entrance lobby between the historical part and the new addition. The main reception for the whole office building and the entries to two lift vestibules are situated here. A tall bright green wall of 30 m high with living plants is the landmark of the atrium together with the onyx reception counter illuminated through as an accent to dark floor tiling in the entrance lobby.

本案最初建造于 19 世纪初。新项目将改造原建筑临街的两翼,并增加一处庭院区。这座历史性大楼的一层由租赁区、咖啡店、餐厅和中庭组成。中庭是由原来的庭院增加一层玻璃屋顶而构成,在原历史建筑和新增加的建筑中间,形成宽敞的入门大厅。整幢写字楼的主要接待处和两个电梯的玄关皆在此处。一座 30 m 高的垂直绿化墙,入口大厅的黑色地砖,明亮的缟玛瑙接待台共同构成了中庭的独特景致。

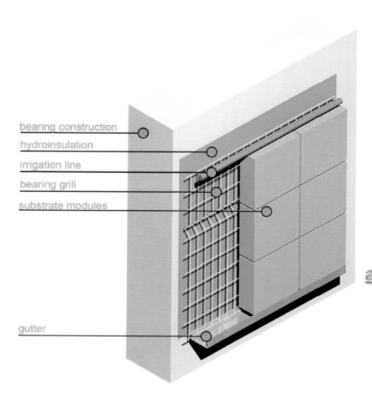

COMPOSITION OF A CONSTRUCTION

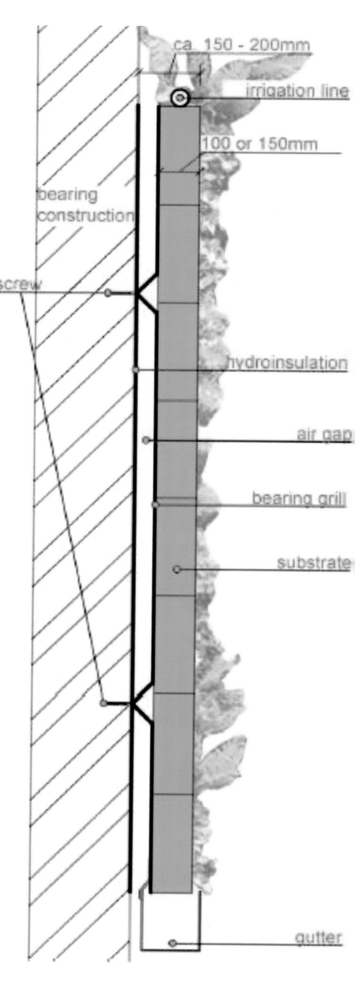

CROSS SECTION

Location
Shenzhen, Guangdong Province, China

Qianhai, Vanke
万科前海

Landscape Architect
Jinhong Greenwall

The Vertical Greenery of Qianhai Enterprise Mansion of Vanke

Floristics : Using the perennial herbs and undershrubs, choosing plants with the color of shades of green, yellow, red as well as various bloomers, matching with plants in the leaf shapes of being round, peach, pointed and linear and shaping with upper-flowing ribbons, the designs give people the senses of being natural and smooth. The plants include Ivy Tree Bark, Kamuning, Dleur-de-lis, Babysbreath, Duranta Repens, Nephrolepidaceae, Rhoeo and gardenia.

Plants for roofs: sedum lineare and callisia.

System Summary: supported with steel structures, module planting system, drop-by-drop irrigation system and micro-spray system.

Design Concept: Working in the office as if in the forest makes people feel natural and wow the world.

万科前海企业公馆立体绿化

植物种类：采用多年生草本和小灌木，颜色上选择深浅绿色、黄色、红色以及各种开花植物，叶形上以圆叶、桃形叶、尖叶、长条形叶等来搭配，设计以向上的飘带来造型，自然、流畅。植物有鸭脚木、九里香、鸢尾、满天星、黄金叶、肾蕨、大蚌兰、栀子花。

屋顶植物：佛甲草、锦竹草。

系统汇总：钢架结构支撑、模块种植系统、滴灌系统、微喷系统。

设计理念：森林办公、自然、令世界动容。

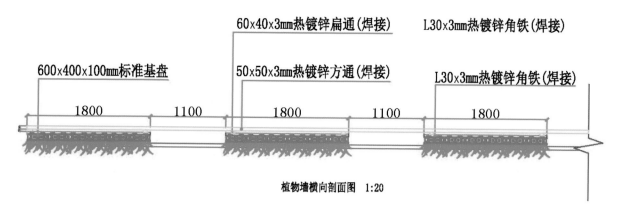

植物墙横向剖面图 1:20

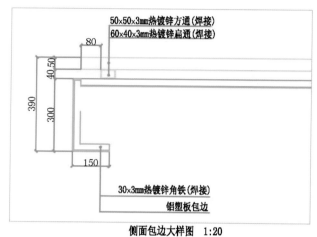

侧面包边大样图 1:20

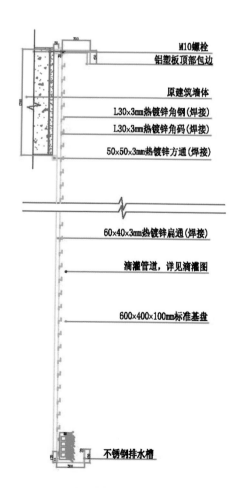

植物墙竖向剖面图 1:15

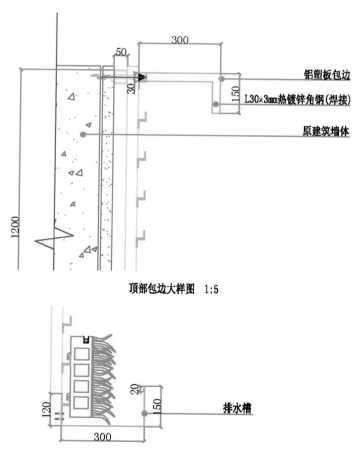

顶部包边大样图 1:5

底部排水槽大样图 1:5

Tianmi Art Space Vertical Greenery

Floristics: Using the perennial herbs and undershrubs, choosing plants with the color of shades of green, yellow, purple as well as various bloomers and with waterfalls as the elements, the designs give people the senses of being natural, magnificent and smooth. The plants include Ivy Tree Bark, Kamuning, Dleur-de-lis, Babysbreath, Duranta Repens, Nephrolepidaceae, Rhoeo and gardenia.

System Summary: supported with steel structures, module planting system, drop-by-drop irrigation system and micro-spray system.

Design Concept: Life on the wall and art on the wall, make you get direct contact with nature while sitting in the office.

天米艺术空间立体绿化

植物种类：采用多年生草本和小灌木，颜色上选择深浅绿色、黄色、紫色以及各种开花植物，设计以瀑布为元素，给人自然、宏伟、流畅的感觉。植物有鸭脚木、九里香、鸢尾、满天星、黄金叶、肾蕨、大蚌兰、栀子花。

系统汇总：钢架结构支撑、模块种植系统、滴灌系统、微喷系统。

设计理念：墙上的生命，墙上的艺术，让微自然与办公室中的您零距离接触！

Location
Shenzhen, Guangdong

Refurbish of TYJ Office Building
深圳桃源居办公楼改造

Designer
Shenzhen Ingameoffice Ltd

Client
LII Industry(Shenzhen) Group Co., Ltd

Client Leader
LI ZHOU, ZHENG TAO

Area
3400 m²

Storey
5 Storeys

Structure
frame - shear wall & local steel

This project is the old office space renovation. In order to break the traditional monotony, repetition and even boring office space, the designers tend to create a space similar to a playful environment to meet the needs of multi-level space of employees. For example, a smaller space sometimes is more conducive to the communication between staff, even more intimate to personal relationships. Each level is unique and interesting designed and manufactured a completely different spatial experience, while connecting the entire office space through a series of lively "cubes". Different with a clear attribution of standard office space, these "cube" spaces are a variety of styles and uncertain programs; they could be reading rooms, snack rooms, gossip spaces, smoking areas, etc. They provide staff with additional kinds of temporary, relaxing and intimate spaces to enjoy, share, or simply be alone.

Setting the green wall system on the west side of the office area resolves two problems. On one hand, plants can enjoy the intense afternoon westward sunlight exposure. By photosynthesis, plants will increase concentration of oxygen anion in the office, which benefits employees' physical and mental health conditions. On the other hand, through the plants shading, the office area is away from western exposure interference, and avoids the formation of Greenhouse effect by western exposure sunlight in the office area, therefore saving additional air-conditioning load.

Moreover, each of the plants is free to be plug in and out from the green-wall. For those staff who like gardening, this feature also allows them to replace their desk plantation whenever they like.

此项目为旧办公楼空间改造。设计打破传统单调、重复甚至冰冷的办公空间，想要创造一种类似于休闲娱乐的空间环境，能够满足员工多层面的空间需求，例如，有时候较小的空间组合反而更有利于员工的交流，甚至建立更亲密的私人关系；每层楼的独特性与趣味性设计在每一层制造完全不同的空间体验，同时将整个办公区内的空间通过一系列活泼的"盒子"串联起来。这些"盒子"空间设置在有明确归属划分的标准办公空间之外，它们风格多样，形式轻松，功能模糊（阅读空间、零食空间、八卦空间、吸烟空间等），为员工提供更多种临时的、放松的乃至更私密的空间来享受、分享或独处。

在办公区的西侧设置的绿墙其实是一举两得。一方面，植物可以尽享下午强烈的西晒阳光，通过光合作用在办公室内提高负离子浓度，有益于员工的身心健康。另一方面，通过植物的遮挡，办公区域免受西晒干扰，避免西晒阳光在办公区形成温室效应，增加空调负荷。

不仅如此，每一个绿墙植株都是可以随意插拔替换的。对于那些喜欢自己养花的员工，此项功能还可以允许他们随时更换办公桌上的植物。

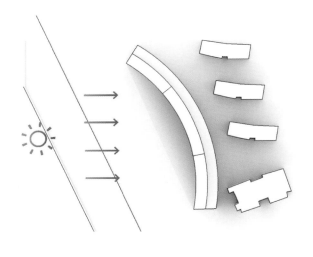
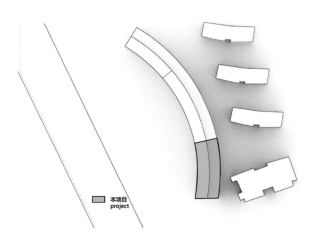

Green wall system is composed of several parallel and equidistant horizontal PVC pipes which function as the water pool. Adjacent horizontal pipes connect to each other through transparent connection tubes at the ends. The system attaches to a steel frame system about 20-meter wide by 20-meter high. Water is supplied from the top of the system by gravity and ensures that all horizontal pipes keep half the pool water level. Excess water is collected in the sump at the bottom of the system, and then pumped up to the tank on the top of the system by a circulation pump. Before the water being recycled into the system, necessary nutrients will be injected into the water automatically. Therefore, a coherent irrigating system is formed which ensures no overflow happens.

Each plant is planted in a recycled plastic water bottle "pot", which is connected with horizontal pipe water pool by a PVC connector. The connectors are about 20cm apart from each other to ensure adequate space to allow plants to grow. A cotton thread connects the pot soil and water in horizontal pipe pool by soaking itself into the water, functioning no different as root capillary action, to ensure water supply of the plants.

The system has the following significant advantages:

1. The translucency: The system can be freely designed into different shapes, double sides planting; two-way access; translucent.

2. Cleaness: The entire system is completely sealed. All plants are irrigated by capillary action, and there is no overflow problem. Particularly suitable for indoor use.

3. Plants replaceable: Because pots are recycled plastic mineral water bottles, and the bottles and PVC water tank are connected by screw threads, so plant replacement is very convenient.

4. Air purification: The large-scale green wall and numerous plants improve indoor air quality significantly; they could increase the amount of indoor oxygen and negative ions.

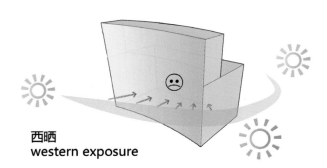

西晒
western exposure

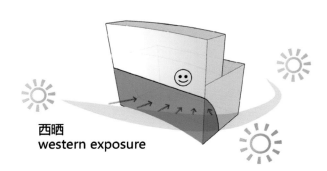

西晒
western exposure

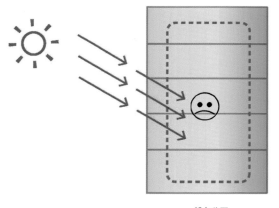

一般办公区
general office area

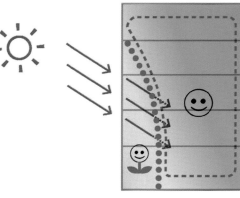

绿墙+办公区
green wall+ office area

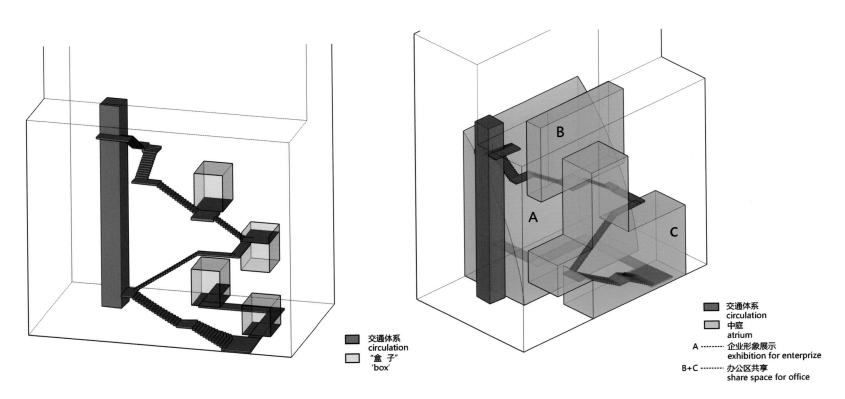

■ 交通体系
circulation
□ "盒子"
'box'

■ 交通体系
circulation
□ 中庭
atrium
A ········ 企业形象展示
exhibition for enterprize
B+C ······ 办公区共享
share space for office

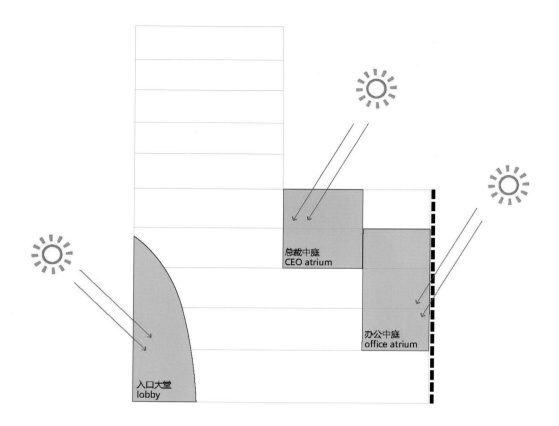

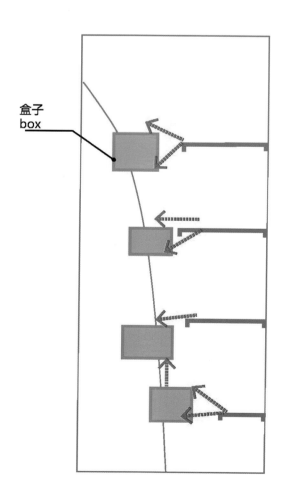

到达系统
access system

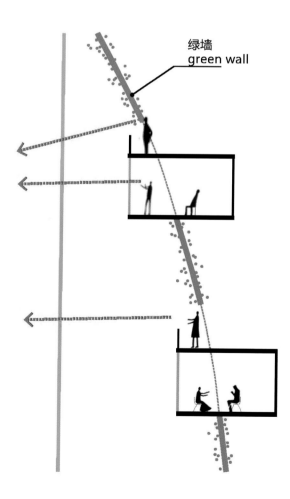

绿墙与盒子系统交错
green wall & box system intersect

　　绿墙系统是由若干个平行而等距的 PVC 横管作为水源池。相邻的水平横管再通过透明连接管彼此相连，再统一附着在一个大约 20 m 宽 ×20 m 高的钢架系统上。水源由系统顶部通过重力流注入，确保所有横管保持半池水位。多余的水源汇集在系统底部的集水池中，再由水泵转移到顶部水箱循环流动。顶部水箱设置一个营养液注入器，定时添加必要的营养。构成一个连贯的，而确保不会溢流的供水系统。

　　每株植物栽种于利用回收矿泉水瓶改制而成的花盆里，通过聚氯乙烯连接头与横管水池相连，每个连接头相隔 20 cm 左右，确保植株之间有足够的生长空间。一根浸泡在水中的棉线将水池中的水引入花盆的土壤，通过毛细作用，保证植物的水分供应。

　　该系统有着以下显著优点：

1. 透光性：系统可随意设计成不同形状，双面种植。具有双向可达透气、透光性。
2. 清洁性：整个系统完全密闭，所有植物利用毛细作用供水，不存在过度浇灌导致水源溢出的问题。特别适合在室内使用。
3. 植株易替换性：由于花盆是回收的矿泉水瓶，瓶口与 PVC 水源池通过螺纹连接。植株替换十分便利。
4. 净化空气功能：由于绿墙规模大，植株众多，可以改善室内空气质量，大幅增加室内氧气与负离子数量。

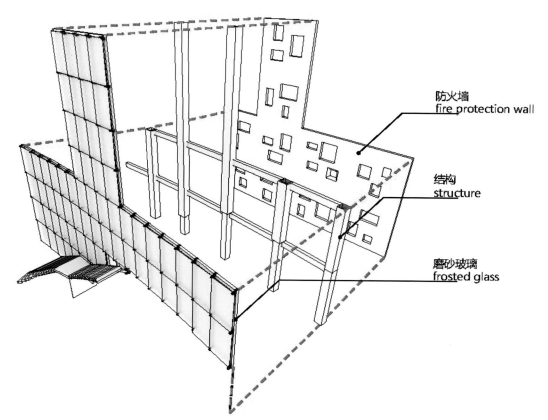

防火墙
fire protection wall

结构
structure

磨砂玻璃
frosted glass

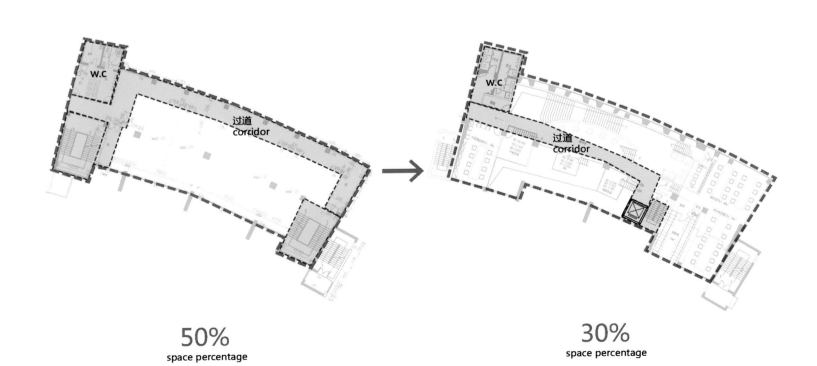

50%
space percentage

30%
space percentage

INDEX
索引

Capella Garcia Architecture

Capella Garcia Architecture is a studio founded in 2001 by the architects Juli Capella and Miquel Garcia, which attempts to pamper the projects, whether large or small. Not being specialists in anything but they are curious about everything, and offer rather than plans, ideas, service and enthusiasm to customers who trust in them. Their passion is architecture and design, in all forms. Their projects include the hotels Omm and Silken Diagonal in Barcelona, Ridaura Auditorium in Santa Cristina d'Aro, Gerona, and recent interior designs for restaurants Jaleo and Minibar by Jose Andres in Washington.

WOHA

The architecture of WOHA, founded by Wong Mun Summ and Richard Hassell in 1994, is notable for its constant evolution and innovation. A profound awareness of local context and tradition is intertwined with an ongoing exploration of contemporary architectural form-making and ideas, thus creating a unique fusion of practicality and invention. WOHA has won an unprecedented amount of architectural awards for a Southeast Asian practice, such as the 2011 RIBA Lubetkin Prize, 2010 International Highrise Award and the 2007 Aga Khan Award for Architecture. Both Directors have lectured at universities in Singapore, Australia, China, the USA, and the United Kingdom, and they have served on various design advisory panels in Singapore.

AgFacadesign

Founded by Kelvin Kan, a registered Architect in the UK and Singapore with more than 20 years experience, he has been actively involved in the glass walls and façade industries for more than a decade prior to starting AgFd. With hands on experience from design to completion in all his projects as an Architect, Glass Specialist and Façade Consultant, Kelvin's passion in designs are found in the details in which he strongly believes makes the difference to how an envelope (or "A"rchitecture) works and lasts.

Leon kluge Garden Design

Leon kluge Garden Design is a international garden design company, concentrating on one of a kind projects that would push the boundaries in landscape design, having done works all over the world such as projects in South Africa, Mozambique, Israel, Comoros, Japan, France, England, USA and New Zealand.
They strive for a healthy mix of architecture, art and interesting, unique and rare plants, that is what make their gardens stand out from the rest.
Their offices are based in South Africa and New Zealand.

Green Studios

Green Studios is a landscape architecture and technology platform specialized in the hydroponic technology applied on green walls and roof gardens.
They have patented in the US a unique type of skin engineered for extremely hot weather along with a smart system. With over 35 projects on our portfolio, Green Studios is considered as one of the pioneers in the region to design, execute and maintain green walls and roof gardens.
Today we are present in Lebanon, Egypt, UAE and the USA. We work with architects, Real estate developers and governmental institutions to change the application of greenery in tomorrow's architecture and influence the field of landscape architecture.

Elmich Pte Ltd

Established in 1985, Elmich Pte Ltd is a leading provider of state-of-the-art and ecologically-minded urban landscaping, waterproofing, drainage, and stormwater management solutions to building developers, contractors, and landscape architects around the world.
Headquartered in Singapore, Elmich's outlook is global, supported by offices in Australia, China, Germany, Switzerland and the United States. Elmich's network of partners span across the globe, covering more than 27 countries.

DLA Landscape and Urban Design

DLA Landscape and Urban Design is a registered practice with the Landscape Institute and a subsidiary within the DLA Design Group. They have a core team of chartered landscape architects with a resource of over 100 individuals throughout the wider DLA Design Group in three offices - Leeds, Manchester and London who provide a comprehensive landscape consultancy service, creating attractive sustainable environments.

Their design philosophy is underpinned by a genuine commitment to truly understand each individual site and its environs. Only by developing an appreciation of the physical and psychological context of a scheme can they begin to develop a design concept that will ultimately resonate with users and enrich the area, both now and for future generations.

Green Empire Industrial Co. Ltd

當代景觀事業有限公司
GREEN EMPIRE INDUSTRIAL CO., LTD

Green Empire Industrial Co., Ltd was set up in 1980s. Under the philosophy of sustainable development, the firm aims to be a full-service landscape architecture firm. The firm has outstanding landscape team to provide professional construction design and planning. With their abundant experience and perspective vision, they act as make-up artist for the land, and coordinate with clients to promote sustainable and eco-friendly development and construction.

MLA (Mission Landscape Architecture)

MLA (Mission Landscape Architecture) is the Landscape Architecture Division of Mission Landscape Companies, a forty year old landscape firm located in Southern California, who in combination with its other divisions, landscape maintenance, tree care, development and environmental resources, provides full landscape services to its clients.

MLA is a Landscape Architecture firm with extensive experience in new ground-up developments as well as renovation and repositioning of existing properties. Services start with development of initial design concepts; continue with refinement of the design ideas, preparation of construction documents and finally construction administration services. MLA's projects span several market segments including retail centers, single and multi-family developments (many with on structure roof top gardens), office complexes, resort and hospitality, and senior living communities in the United States and beyond. Their clients are among the "Who's Who" of the development world.

CTLU Architect & Associates

CTLU Architect & Associates was founded at the intersection of creative sensibility and rational practice.

They are committed to all kinds of spatial design, pursuing a better spatial quality. They integrate the ideas of sustainable green building, to create a healthy and comfortable environment. They firmly believe that the creative thinking and practical function can coexist, and persevere in the best quality without prejudice to efficiency.

They offer quality design and service to their client and also provide a fast-growing, both widely and deeply learning environment to their colleagues.

DaM spol. s r.o.

- Founded in 1989
- Partners - architects: Petr Burian, Richard Dolezal, Jiri Havrda, Jiri Hejda, Jan Holna, Petr Malinsky
- 20 employees
- Private projects and housing (including villa for former President of the Czech republic Vaclav Havel)
- Reconstructions of important historic buildings (including buildings at Prague Castle, palace of Boscolo Hotel)
- Modern buildings in strategic places (Euro Palace on Wenceslas Square Prague, Filadelfie Office Building, Prague, Main Point Karlin Prague)
- Interiors and exhibitions (The Story of Prague Castle, Pilsner Urquel permanent exhibition)

In 2009 DaM spol. s r.o. performed the 20 years of its activity by large exhibition in Fragner Gallery Praha and published the book "DAM ARCHITECTS". Realizations of DaM Architects received many awards, among the most important:
- Several Grand Prix of Architects of The Czech republic
- Several Best of Realty of The Czech republic
- Nominated for the Mies van der Rohe
- MIPIM Award

ASPECT Studios

ASPECT Studios™ - ASPECT Studios are a group of Landscape Architects who design places where people want to be, providing the best for landscape architecture, urban design, high-end interactive digital media and environmental graphics.

With six studios located within Australia and China, ASPECT Studios is an Australian owned business with an award winning industry leading track record both for the projects that they design and also for the way that they think.

STGK

STGK is a multiple design studio, founded by Gen Kumagai in 2009. Prior to establish his own design studio, he worked for STUDIO Choi, Jae-Euna as a senior designer and an associate at earthscape inc. He is currently an adjunct professor at the Aichi University of the Arts. The studio mainly engages in Landscape design, Product design, Public Art and City planning. They have won the Good Design Award 2009/2010, and more recently the Nominative Competition for Yokohama-Station West Gate Project, which will be completed in 2020.

FAAB Architektura

FAAB Architektura was founded in 2003 by Adam Bialobrzeski and Adam Figurski after 5 years of professional experience working in architectural practices both in Poland and abroad. Maria Messina joined the design team in 2005 having had 7 years prior professional architectural training in both the USA & Poland. Our team is composed of professional architects and engineers with over 10 years' experience in the field of architectural practice and building construction.

FAAB Architektura maintains design excellence by participating in national & international architectural design competitions gaining awards and recognition in over eleven of such entries. Three of these winning projects have been successfully constructed – namely: PGE GiEK CONCERN HEADQUARTERS in Belchatow, REGIONAL BLOOD CENTER in Raciborz and the FOUNDATION FOR POLISH SCIENCE in Warsaw.

DP Architects

Founded in 1967, DP Architects was one of the firms responsible for the urban landscape of Singapore. Now a leading architectural practice in Asia with over 1,200 staff and 15 offices worldwide, the firm provides a range of services from architecture, urban planning, landscape design, infrastructure design, engineering, sustainable design and interior design to project management. DP Architects has a long history in a wide variety of projects with a particular expertise in vast undertakings such as Suntec City, Esplanade-Theatres on the Bay and The Dubai Mall. Current landmark projects include Resorts World Sentosa and Singapore Sports Hub. The firm was founded with a deep concern for the built environment and the need to create architecture of excellence that enriches the human experience and spirit.

Grant Associates

Grant Associates is a British Landscape Architecture consultancy specialising in creative, visionary design of both urban and rural environments worldwide, working with some of the world's leading architects and designers.

Inspired by the connection between people and nature Grant Associates fuses nature and technology in imaginative ways to create cutting edge design built around a concern for the social and environmental quality of life.

Grant Associates has experience in all scales and types of ecological and landscape development including strategic landscape planning, master planning, urban design and regeneration and landscape for housing, education, sport, recreation, visitor attractions and commerce.

SAA Architects

SAA has over 40 years of architectural and design expertise in Singapore and its region. Established in Singapore in 1970, the firm has been playing a key role in shaping the Singapore built-environment. SAA has won a multitude of accolades through the years of practice. Being recognized as one of the architectural firms that has the greatest impact on the built environment in Southeast Asia, the firm has won BCI's Asia Top Ten Architects in Singapore Awards consecutively from 2009 to 2012 and in 2014. The firm was also awarded with the FIABCI Singapore Property Awards for real estate design excellence in 2012. SAA designed and realised several sustainable projects with Greenmark Platinum Awards and Universal Design (UD) Mark Gold Plus Award – the Continuing Education and Training (CET) Campus in Paya Lebar, Jem in the Jurong Gateway District, and One Raffles Place Tower 2 in the Central Business District.

SuperLimão Studio

SuperLimão Studio started its activity in the end of 2002. They believe that the concept of quality of life is closely related to the balance between body, mind and soul. Thus they seek to understand the meanings of spaces and symbols, to interpret and materialize it resulting in spaces. Instigate a smile is part of our daily lives. They love their work and they don't separate their personal lives in a parallel universe. A critical analysis is present and constant in their daily lives. Always looking to improve what they see or feel in order to achieve excellent results.

Jinhong Greenwall

Jinhong Greenwall is an urban vertical landscape system operator of research and development, design, production, construction and maintenance. The business section includes vertical landscaping, roof landscaping, slope protection landscaping, bridge landscaping and creative greenery products.

Jinhong Greenwall's green systems are produced with standardization, scale and industrialization. They are easy to construct and maintenance, with creative and elegant look, sustainable and energy efficient features.

Jingong Greenwall has been partners with well-known enterprises in China, such as Shenzhen Urban Management Bureau, Vanke, Greenland, Tencent, Country Garden Holdings, Gragon Group, Building Research Institute, Rongqiao Group, etc. The projects spread over Shenzhen, Guangzhou, Changsha, Chengdu, Shanghai, Fujian, Hainan, etc.

Fytogreen Australia

Fytogreen Australia Pty Ltd was founded in February 2002, when the company became the established licensee in Australia, following on from 23 years of product development by Fytogreen originators in Europe.

Over the past 13 years, since establishment in Australia, Fytogreen has become an acknowledged industry leader and innovator in horticultural technologies. They are a research focussed, design and construct supplier of vertical gardens, roof gardens and green facades and all of Fytogreen's design and project management team are university graduated.

Fytogreen have completed more than 150 successful vertical garden projects in Australia and in international locations, including California, Singapore and Dubai and supplied proprietary roof garden media components to approximately 320,000 m² of roof gardens throughout Australia.

Chartier-Corbasson Architecte

In 2013, the Chartier-Corbasson architects agency received a prestigious award, the "Prix Dejean". This award greets the web that these two architects has spun, characterized by a relevant and consistent work, regularly winning in France and abroad, the pace of a project every two years on average.

Their projects are consecutive but all different, thought there is something telluric in their way to make the buildings emerge from the ground. That draws a thread through this work in progress, and gives an undeniable force their choices.

Tenacity, consistency and commitment are words that characterize these two, as well as their appetite for research and technical innovation, which focuses on forms or on materials, and their obvious pleasure to share with entrepreneurs without which nothing would be.

Pomeroy Studio

Pomeroy Studio is an award-winning international team of designers and thought leaders of sustainable built environments. The studio comprises of master planners, landscape architects, architects, interior and graphic designers, as well as sustainability consultants and academics. Quantitative and qualitative research complements an interdisciplinary design process that lies at the foundation of their creative design and decision-making. This has allowed the studio to generate people-centred places, from the micro-scale of dwellings to the macro-scale of cities, that pushes the envelope of design and research by balancing a 'creative vigour with an academic rigour'.

Kono Designs

Founded in January 2000 by Yoshimi Kono, Kono Designs is a New York based multi-disciplinary practice active in architecture, interior design, corporate identity, graphics and product design. With clients based in Asia, Europe and USA, the practice handles wide ranging project from high-rise towers to wrist watches.

By methodically arranging a series of goals and complex design requirements, Kono Designs communicates concept and function through tangible forms and creating works that are clean and unobtrusive.

Art & Build Architects

Since 1989 Art & Build have made creativity their trademark, individual and collective fulfillment their philosophy, and know-how their reputation.

As a European architecture practice located in Brussels, Paris, Luxembourg and Toulouse, the Art & Build team is active across both public and private construction sectors, from office to industrial, healthcare to residential, retail and leisure, culture and education, and from urbanism to landscape design.

Building the cities of tomorrow within the economic, environmental, social and cultural climate is an ongoing goal to which Art & Build responds through innovation and foresight.

Greenology

Greenology is a team of passionate botanists, horticulturists, engineers, designers and nature lovers, driven by green conscience to green their world, one wall at a time.

The company designs and create sustainable living systems – building green walls for urban ecosystems to remain productive and flourish, over time. Plant species and substrates have been carefully researched in developing their proprietary products such as GNanoFibre™ and GMatrix™, which ensure sustainable plant growth and development, hence reducing plant mortality and replacement rates. Greenology strives to not only provide healthy green walls upon installation, but for them to remain healthy as well.

VenhoevenCS architecture+urbanism

VenhoevenCS architecture+urbanism is an innovative office for sustainable architecture, urban planning and infrastructure, specialized in finding integral, spatial solutions for social and cultural issues, on every possible scale. VenhoevenCS architecture+urbanism believes cities and public spaces can become attractive, productive and flourishing meeting places by applying cultural, technical and spatial design innovations. In spatial and cultural sense, a VenhoevenCS design (either for a city or building) is a patchwork of different (sub) cultures, a lively cosmopolitan whole and an ideal biotope for pedestrians. By combining urban functions in high density, space remains for interesting interaction and green areas. Squares, parks and courtyards are the cities lungs and give opportunity for meeting friends, playing sports and organising events.

Oscar Gonzalez Moix

Oscar Gonzalez Moix, was born in Buenos Aires Argentina and is formed as an architect at the University of Buenos Aires (UBA), Faculty of Architecture, Design and Urbanism, Argentina.

In 1998 he founded his architectural firm in Buenos Aires, running projects, primarily single-family, multifamily real estate development and management. After some projects confirmed for Peru, decided to install its main studio in Lima, in 2002, solving a variety of programs and scales.

His interest in experimentation and new models, as well as his deep knowledge of the construction, define an architecture that combines tradition and innovation modern contemporary, simple, strong, sensitive and friendly place with who lives there. Participate in academic teaching, closes the circle of passion for architecture, a conceptual reflection that proves its success.

Ciel Rouge Creation

Ciel Rouge Creation is a French-Japanese established architectural office based in Tokyo and Paris. Between the East and West, ideas are exchanging from one side to another. These complementary cultures are enriching themselves in design and architectural concepts. French architect, Henri Gueydan, the founder of Ciel Rouge Creation is involved in adapting architecture to its most precise function. The space should be the largest space possible, even on a very small scale. The idea is to create the ideal "weight of life" by smooth shapes, well positioned perspectives, measured lightness, adapted materials and colors. The space could be discrete, elastic, convenient and powerful at the same time. Through this philosophy, the construction has to be simple, appropriate, and well adapted to improve the "taste of existence".

AECOM

AECOM is a premier, fully integrated professional and technical services firm positioned to design, build, finance and operate infrastructure assets around the world for public – and private-sector clients. With nearly 100,000 employees – including architects, engineers, designers, planners, scientists and management and construction services professionals – serving clients in over 150 countries around the world, AECOM is ranked as the #1 engineering design firm by revenue in Engineering News-Record magazine's annual industry rankings, and has been recognized by Fortune magazine as a World's Most Admired Company. AECOM provides a blend of global reach, local knowledge, innovation and technical excellence in delivering customized and creative solutions that meet the needs of clients' projects.

EZEQUIELFARCA architecture & design

Ezequiel Farca is Creative Director and CEO of the studio EZEQUIELFARCA architecture & design. In 1995, Ezequiel opened his first design studio, and shortly afterwards a furniture showroom in the heart of the Polanco district in Mexico City where he has set up his studio practice, which comprises a team of 30 specialists in architecture and interior design.

With over 24 years of experience, Ezequiel Farca has an outstanding profile having begun his career designing high-quality and "custom-made" furniture, before breaking out and focusing on interior design. With his experience at different levels of design, today he has become a talented architect, sought after for his attention to detail and good taste, since everything he designs—whether an item of furniture, a home or a hotel—he does so with the same precision. Now the firm is working around Mexico, L.A. and Italy.

Stanley Beaman & Sears

The principals of Stanley Beaman & Sears have a shared vision that has developed over the course of a 30-year friendship and professional association. With a collaborative working style and an appetite for innovation, they started the firm to focus on unique or pioneering projects in healthcare, higher education and the arts–building types that, in their view, represented "civilization's great institutions".

Stanley Beaman & Sears is unique in that the principals work collaboratively with their staff on each project from start to finish, focusing on function, form and technical issues respectively. Their level of personal involvement is unique in the industry.

Tierra Design (S) Pte Ltd

Tierra Design (S) Pte Ltd is a multidisciplinary design firm established in 1995 in Singapore by Franklin Po. As designers with a focused desire to explore the relationship of the arts, culture, nature and land, Tierra provides professional design services in architecture, landscape architecture, master planning, urban design and sustainable environmental design, in order to enhance the quality of life for people who occupy the spaces that we create.

Tierra's projects are known for its economy of planting and sensual natural simplicity, beauty and quiet reserve. Tierra has been the recipient of a number of awards for design excellence. Its projects include prestigious residential, commercial, hospitality and institutional projects in Singapore, Malaysia, Indonesia, India, China and Abu Dhabi. Tierra embraces diversity in its people, with collaboration playing a vital role in creating unique, thoughtful and holistic designs. Tierra's aim is to help clients make intelligent and well-informed decisions to enhance value for them and create a better future for our planet.

ACKNOWLEDGEMENT 鸣谢

We would like to thank everyone involved in the production of this book, especially all the artists, designers, architects and photographers, for their kind permission to publish their works. We are also very grateful to many other people whose names do not appear on the credits but who have provided assistance and support, for their contribution of images, ideas and concepts, as well as their creativity to be shared with readers around the world.

在此，我们非常感谢参与本书编写的所有人员，尤其是各位艺术家、设计师、建筑师和摄影师，他们授权我们出版他们的作品。我们亦感谢为本书编写提供支持和帮助的人员，虽然他们的名字并没有出现在本书中，感谢他们提供的图片、思想和理念，并将他们的创意分享给全世界的读者。

The people involved in the book are:

参与本书编写的人员有：

Veera Sekaran, Michael Leong, SAA Architects, Greenology, WOHA, Patrick Bingham-Hall, Salad Dressing, Elmich Pte Ltd, DP Architects Pte Ltd, Cicada Pte. Ltd., Grant Associates, Tierra Design (S) Pte Ltd, SOM, AgFacadesign, ICN Design International Pte Ltd, Hassell, Robert Such, Ciel Rouge Creation, Adam Bialobrzeski, Adam Figurski, Maria Messina, Doraco Construction, Fujita Corporation, STGK, Takahiro Shimizu, Daiwa Lease, Kono Designs, Green Empire Industrial Co., Ltd, CTLU Architect & Associates, Junichi Inada, WIN Landscape Planning & Design Pte Ltd, JST Malaysia, Kaori Ichikawa, Nakano Construction sdn Bhd, Fytogreen Australia, CRCI de Picardie, R.Meffre & Y.Marchand, Chartier-Corbasson Architectes, VenhoevenCS architecture+urbanism, Broadway Malyan, Architects 61 Pte Ltd, Amir Sultan, Jerde Partnership Inc., Pomeroy Studio Pte Ltd, Tom Epperson for Century Properties, DLA Landscape & Urban Design, GF Tomlinson, ANS Group, Lowe Engineering Ltd, Leon Kluge, CNBV architects, Green Studios, SOLIDERE, Fountain Workshop, Spiers and Major, AECOM, Gonzalez Moix Arquitectura, Antonhy Isles Infanta, Juan Solano Ojasi, Virginia Burt, CSLA, ASLA, Visionscapes, Landscape Architects Inc., Stanley Beaman & Sears, Hanson PhotoGraphic, SuperLimão Studio, Studio Campana, EZEQUIELFARCA architecture & design, Jaime Navarro, Roland Halbe, Puerto Vallarta, Art & Build Architects, P. LEBLANC & P. SPEHAR, MLA (Mission Landscape Architecture), Elenberg Fraser Architects, Tract, Lu Yanni, DAM Architects, m3m, IMMOFINANZ Group, Arch Vegetal, Filip Slapal, Jinhong Greenwall, etc.